INSIGHTS ON SCIENCE JOURNALISM

Bringing together experts from a range of disciplines, this collection critically examines science journalism, paying special attention to the points of tension that science journalists navigate in their work today.

Faced with the twin crises of climate change and a global pandemic, science journalism has never before been so prominent. This book showcases perspectives that transcend the particulars of the specific news events and outlets studied, in order to provide an overview of the key areas of scholarly interest regarding the nature of science journalism. The volume is organised into three sections: the first provides historical case studies illustrating the demarcation of science journalism from science as science journalism emerged as a recognisable news beat in the twentieth century; the second examines the relationship between science journalists and their sources, particularly scientists, and the mediation of this relationship through organisations, foreign journalism and political constraints; and the final section considers the style and voice of science journalism content. Case studies and original empirical research are compiled from across the globe, including the UK, US, Germany, Vietnam, and Russia, and are synthesised to offer a readable and engaging insight into the beat.

Insights on Science Journalism is recommended reading for advanced students and researchers of science journalism and communication and will also appeal to those working in the fields of science and technology studies and risk communication.

Felicity Mellor is Director of Science Communication Unit at Imperial College London, UK, where she oversees the Unit's long-running masters programmes in science communication. Her research focuses on science journalism and the ideological dimensions of media discourse about science. Her publications include two co-edited books: *The Silences of Science* and *Science and its Publics.*

Journalism Insights

The Journalism Insights series provides edited collections of theoretically grounded case study analyses on an eclectic range of journalistic areas, from peace and conflict reporting to fashion and sports reporting.

The series has a bias towards the contemporary, but each volume includes an important historical, contextualising section. Volumes offer international coverage and focus on both mainstream and 'alternative' media, always considering the impact of social media in the various fields.

The volumes are aimed at both undergraduate and postgraduate students on journalism as well as media and communication programmes who will find the texts original, interesting and inspirational.

For information on submitting a proposal for the series, please contact the Series Editor Richard Lance Keeble, of the University of Lincoln and Liverpool Hope University, at RKeeble@lincoln.ac.uk

Insights on Fashion Journalism
Edited by Rosie Findlay and Johannes Reponen

Insights on Immersive Journalism
Edited by Ana Luisa Sánchez Laws

Insights on Science Journalism
Edited by Felicity Mellor

For more information visit: https://www.routledge.com/Journalism-Insights/book-series/JI

INSIGHTS ON SCIENCE JOURNALISM

Edited by Felicity Mellor

Routledge
Taylor & Francis Group
LONDON AND NEW YORK

Designed cover image: © Getty Images

First published 2024
by Routledge
4 Park Square, Milton Park, Abingdon, Oxon OX14 4RN

and by Routledge
605 Third Avenue, New York, NY 10158

Routledge is an imprint of the Taylor & Francis Group, an informa business

British Library Cataloguing-in-Publication Data
A catalogue record for this book is available from the British Library

ISBN: 978-1-032-35062-2 (hbk)
ISBN: 978-1-032-34128-6 (pbk)
ISBN: 978-1-003-32672-4 (ebk)

DOI: 10.4324/9781003326724

Typeset in ITC Galliard Pro
by KnowledgeWorks Global Ltd.

CONTENTS

CONTRIBUTORS

Alexandra Borissova Saleh is a science communication consultant at E-Quadrat Science & Education in Berlin, Germany. She is a former director of the MSc in Science Communication at ITMO University in Saint Petersburg and the first president of the Russian Association for Science Communication (AKSON). She has more than ten years of experience as a science journalist and has undertaken the first substantial investigation into Russian science communication in a project supported by the Alexander von Humboldt Foundation's German Chancellor Fellowship.

Irene Broer is a researcher at the Leibniz Institute for Media Research, Hans Bredow Institute in Hamburg, Germany, where she studies new intermediaries in science communication. With a background in cultural anthropology and journalism studies, she is interested in the ways that editorial practices, norms and ideologies shape the ways that scientific knowledge and expertise are mediated in public communication.

Jane Gregory is a researcher and teacher of science communication in the UK. She has designed and delivered award-bearing programmes for Imperial College London, University College London, the University of Manchester and, most recently, the University of Cambridge. Her research focuses on the twentieth-century and recent history of popular science. She is also involved with third-sector organisations, in public science engagement and science policy.

Lars Guenther is Professor of Communication Science at LMU Munich's Department of Media and Communication in Germany, and Extraordinary

Associate Professor at the Centre for Research on Evaluation, Science and Technology (CREST) at Stellenbosch University, South Africa. His research interests focus on science and health journalism, as well as public trust and communication of risks and (un)certainty.

Jared Robert Keller is a researcher at the Open Data Institute in London, specialising in evidence-based foresight, science communication and the history of technology. His doctoral research examined the post-war professionalisation of science journalism on BBC radio.

Lauren Kilian is Director of the Office of Disability Services at Marymount Manhattan College in New York City, USA. She holds a PhD in English from Stony Brook University, where her research explored literary intervention in scientific debates and the role of narrative in public science discourse. She has taught literature and writing composition and has developed a study on the impact of habitual writing on undergraduate students' relationship to their writing.

Felicity Mellor is Director of the Science Communication Unit at Imperial College London, UK, where she oversees the Unit's long-running masters programmes in science communication. Her research focuses on science journalism and the ideological dimensions of media discourse about science. Her publications include two co-edited books: *The Silences of Science* and *Science and its Publics.*

Katarzyna Molek-Kozakowska is Associate Professor and Head of the Department of English at the University of Opole, Poland, and Senior Research Fellow at Vilnius Gediminas Technical University, Lithuania. With a background in English studies, she specialises in discourse analysis, science communication and media studies. She co-edits *Res Rhetorica*, an international open-access journal focused on rhetoric and linguistics.

An Nguyen is Professor of Journalism and Co-Director of the Centre for Science, Health and Data Communication Research at Bournemouth University, UK. He began his career as a science journalist in Vietnam. His books include *News, Numbers and Public Opinion in a Data-Driven World* and *Developing News: Global Journalism and the Coverage of 'Third World' Development.*

Minh Tran is Lecturer in Journalism and Communication and Deputy Head of the Department of Linguistics, Literature and Journalism at Danang University of Education, Vietnam. She holds a PhD from Bournemouth University in the

UK, in which she examined the coverage of global science controversies in Vietnamese news journalism.

Anna Tratter graduated in 2021 from the Ludwig Maximilian University of Munich, Germany, and in 2022 from the Technical University of Munich, Germany. In her BA thesis, she focused on press releases of the German Chemical Society (Gesellschaft Deutscher Chemiker). Besides her research interests in science journalism, she has worked both in the chemistry lab and as a science journalist. In 2023, she became a student at the German School of Journalism in Munich.

Justin T. Schröder is a Research Associate at LMU Munich and the University of Hamburg, primarily working on the project 'The trust relationship between science and digitized publics' (TruSDi), funded by the German Research Foundation (DFG). In addition, he is a PhD student at Universität Hamburg, Germany. His research interests include science, political and digital communication.

Susan E. Swanberg is Associate Professor of Journalism at the University of Arizona School of Journalism in the US, where she teaches news reporting, science journalism, environmental journalism and media law. She has also worked as a criminal lawyer, a bench scientist and a journalist.

INTRODUCTION

Beating the bounds of science journalism

Felicity Mellor

An ancient custom in English villages, known as beating the bounds, consists of villagers walking the boundary of their parish as a reminder of where their lands begin so that they can prevent encroachment by neighbouring villages. The 'beat' of a journalist derives from the same idea of traversing familiar territory through habitual visits, as when the crime reporter attends the court in a newsgathering routine that similarly aids social learning and remembering. The metaphor is particularly apt in the case of the subject of this book: science journalism not only involves visiting the same news sources each week (in this case, the press releases of the most prestigious science journals) but to become established as a newsroom specialism has had to maintain boundaries with other journalists, with other types of communicators and with scientists.

Science journalists have sometimes been keen to downplay the boundaries within journalism in order to play up their identity as journalists. As a science editor at a UK daily newspaper once said of his specialism: "I don't believe that it exists, there is journalism which happens to be about science, just as there is journalism which happens to be about football or golf or whatever, I don't believe that there is a specialist area" (Hansen, 1994: p. 116). Yet, the importance of general journalistic skills notwithstanding, science journalism is a recognised beat with designated reporters and editors at major news outlets around the world, professional organisations and distinctive approaches to news gathering and story treatment.

Historically, the English villagers beating the bounds would carry sticks with which to thrash the boundary markers as they processed around their parish. To further reinforce the memory, children would sometimes undergo a trial at prominent landmarks – such as having their heads knocked against

DOI: 10.4324/9781003326724-1

a boundary stone (Soth, 2020). It takes the metaphor only a little too far to suggest that this also resonates with science journalism. While physical pain may not be involved in this case, science journalism is frequently subjected to the criticisms of others wielding greater epistemic authority – the scientists on whom the journalists report. The science journalism beat has frequently been the subject of a beating.

A large and growing body of scholarly work has considered the validity of such criticisms. Much of this work takes a particular area of science as the starting point and examines questions such as whether the coverage replicates scientific discourse, to what extent it is enfolded within political agendas and how it contributes to new cultural associations and meanings. By focusing on particular topics within science, such studies often implicitly start from the scientists' perspective rather than from newsroom experiences or the wider cultural context. Other studies have interviewed science journalists to reveal the norms and values that shape their output or have examined all the science coverage of chosen outlets in order to capture the full breadth of science-based news, much of which is not produced by specialist beat reporters.

Despite the considerable size and scope of this body of scholarship, few works have attempted a historical overview of science journalism as a specialism distinct from other forms of popular science; Dorothy Nelkin's (1987) account of the development of US science journalism across half a century remains a classic in this respect. Having first emerged as a specialist beat in the 1920s in the United States and the United Kingdom, science journalists' place in the newsroom was secured in the following decades by the increasing orientation of society towards technocratic policies and by dramatic technological developments, such as atomic weapons and the space race. In other countries, the recognition of science journalism as a specialist beat came later, depending not only on local newsroom cultures and the structure and ownership of the news media, but also on wider socio-economic factors such as the form of government, the perceived role of science in economic development and the presence of national science organisations.

Where science journalism ends and related beats – such as medicine, development, the environment, technology, energy or agriculture – begin has evolved over time and depends on the size of the newsroom and editors' perceptions of audience interests. One re-drawing of the boundaries of the beat occurred in response to the environmental movement of the 1960s and 1970s, when some outlets appointed new specialists to cover environmental issues, often with more of an advocacy approach than was the norm among science journalists (Fahy, 2017). Science journalists, by contrast, retained an affinity with the aims of other popularisers of science through much of the past century: encouraging public interest in science by explaining the relevance of research findings.

In the past two decades, changes in media technology have once again broken open the boundaries of the beat as part of the widescale challenges impacting all of the news industry. Online popular science outlets have benefitted from the narrowcasting made possible by targeted advertising and scientists and their institutions are now able to reach audiences directly without the mediation of a news outlet. Increasing job opportunities for professional science communicators at scientific organisations and decreasing job security for staff journalists at news outlets have combined to lead to a diversification of careers for those producing science journalism. There are two contrasting ways of interpreting these changes: one tells the story of science journalism as a field of opportunity; the other is a story of perpetual crisis. The more prominent version is the story of crisis.

A story of crisis

News journalism, the crisis account tells us, struggles to accommodate the features of science that make it a uniquely reliable source of knowledge, with the result that science is either ignored or distorted in the news media. Throughout the past century, scientists – and sometimes news producers – have complained about the poor quality of the news coverage of science. Some US and UK examples from twenty-year intervals illustrate the recurring nature of these complaints. Take, for instance, US news magnate E.W. Scripps, 1919 complaint that every scientifically minded newspaper reader "is constantly reminded that there is a vast quantity of misinformation being constantly spread abroad by our newspapers" (Scripps, 1919: p. 306). Two decades later, in 1940, German-American science writer Willy Ley complained that it was "woefully hard to convince newspaper people that they should open some space for science" (Buss, 2014: p. 208). Another two decades later, in 1961, the science journal *Nature* identified in the British Press "a deplorable lack among some writers of a true appreciation of the technique, aims and objectives of science" (Nature, 1961: p. 292). In 1985, a report by the UK's Royal Society reached a similarly despairing conclusion: "The whole process of reduction of the scientific information to a manageable form by the science journalist is almost bound to lead to some distortion" (Bodmer, 1985: p. 21). A British medic and opinion columnist voiced the same concerns in blunter language twenty years later, when he asked: "Why is science in the media so often pointless, simplistic, boring or just plain wrong?" (Goldacre, 2005).

The constancy of these concerns suggests that they transcend assessments of journalists' output and serve to enact a power dynamic, primarily between scientists and news producers, but also between science and the state. In Western democracies, science journalism became established as a beat as scientists attempted to assert an essential role for science in social progress (see Chapter 1). The later emergence of science journalism in the countries of the Global

South was similarly coupled to ideas about the role of science in economic development. The recurring complaints about science journalism are premised on assumptions about the role of science in society, who should be in control of the public image of science and who has the competency to communicate science accurately and meaningfully. These play out through two differing, sometimes conflicting, sets of professional expectations – journalistic and scientific – with the degree of journalistic autonomy shifting over time and depending on the topic at hand. These differing expectations result in a recurring sense of crisis.

Science journalists themselves more often identify a crisis stemming from their current working conditions. Typically, this is presented as a narrative of decline: things are getting worse. A 2021 survey of over six hundred science journalists worldwide showed an increase in the proportion agreeing that science journalism was in crisis: only a third of respondents rejected the idea, down from almost half a decade earlier. Fewer than half of these journalists reported being happy with their jobs and a majority thought that editors were uninterested in science stories.

A key moment in this narrative is the closure of the entire CNN science and environment desk in 2008 (Brainard, 2008). A global survey of almost 500 science journalists in 2009 found that a third of science journalists had experienced cuts to the beat at their workplaces in the United States and Canada and about a quarter had in other countries (Brumfiel, 2009). One science journalist interviewed in 2016 for an Australian study described science journalists in that country as "an endangered species" (McKinnon et al., 2018: p. 568). For those who remain, the job is precarious. By 2018, nearly half of the membership of America's National Association of Science Writers were freelances (Dunwoody, 2021). Nineteen science journalists interviewed for a study in 2022 all identified a decrease in stable employment and a rise in freelancing as the major challenges the field faced in the United States (Anderson and Dudo, 2023).

Those who remain as staff reporters face increasing workloads as they work across multiple platforms in a 24/7 rolling news environment and increased stress as their output is monitored by their editors for the number of clicks they receive. In the United Kingdom, a science reporter at an online tabloid will typically write four to six articles a day. In this context, shortcuts presented in the form of public relations materials become highly attractive. About seven in ten respondents to the 2021 global survey felt that PR was driving science news and that the proliferation of press releases combined with cost cutting in newsrooms was leading to poor quality reporting (Massarani et al., 2021, pp. 27-28).

This last fits with a narrative of decline that is also commonly found in academic studies examining the dependency of science journalism on PR. As discussed in Chapter 4, a significant proportion of science news is derived, with little or no alteration, from press releases. Science PR operations have always had a significant influence on science journalism, but in recent decades, scientific journals, professional societies and universities have all greatly enlarged their communications offices; some evidence suggests that the PR-dependency of science is increasing as a consequence (Vogler and Schäfer, 2020). The reproduction of PR materials means that the framing of science news is frequently determined by the interests of scientific institutions rather than the public interest, encouraging what science communication scholars Martin Bauer and Jane Gregory have characterised as a shift in "the logic of journalism towards a logic of corporate communication" (Bauer and Gregory, 2008: p. 33). Only when science stories have far-reaching and contested policy implications does science reporting take a more independent stance.

As journalism scholar Sharon Dunwoody (2001) suggested, the boundary between science and PR is blurred further if freelance journalists spend some of their time writing publicity materials for scientific institutions. Over a third of respondents in a 2016 survey of members of the US National Association of Science Writers described themselves as press relations professionals, half of whom also worked as journalists (NASW, 2016). One function of journalism is to scrutinise institutions that have the capacity to affect the lives of ordinary people. That function is compromised if journalists are dependent on those institutions for paid work.

Another major pressure comes from the opposite direction: the circulation of science-related misinformation by new media actors who have no connections to either the institutions of science or those of professional journalism. The disruption of the gatekeeping function of the legacy news media brought about by social media has enabled content producers with no alignment to traditional journalistic values to reach large audiences in order to advance specific agendas. This challenge came to the fore during the Covid-19 pandemic, when UNESCO declared that a 'disinfodemic' was running alongside the disease pandemic (UN, 2020); it has also been a long-standing issue with reporting climate change. As a result, professional science journalism may be losing trust among some portions of the public and science journalists can feel under constant attack on social media, causing emotional burnout (Anderson and Dudo, 2023).

The Covid pandemic also serves as a reminder of the physical dangers that science journalists can face when reporting on science-based policy issues in media environments with little or no tradition of a free press (see Chapter 7 for the case of Russia). Journalists in such regimes face arrest and imprisonment if they fail to comply with government censorship. According to Reporters Without Borders, 135 people were arrested around the world during 2020 for their reporting of the pandemic (RSF, 2020: p. 11).

A story of opportunity

Positive accounts of the current state of science journalism are far less common than the story of crisis. Yet, as well as posing severe challenges, the Covid pandemic has also served to highlight the importance and reliability of professional science journalism. For instance, in a notably optimistic blog in which he compared science journalists to superheroes, Al-Jazeera medical journalist Osama Abu El Rub argued that the pandemic had demonstrated that, more than ever, the world needed "facts-based science and medical journalism". With proper investment, El Rub proposed, "science journalists will save the world" (Abu El Rub, 2022).

Writing before the pandemic, US journalists and science communication lecturers Thomas Hayden and Erika Check Hayden were similarly optimistic, declaring: "We live in a golden age of science and environmental journalism" (2018: p. 1). While noting that careers are precarious, they suggested that "there has never been more, better quality science and environmental journalism produced than there is today" (p. 1). According to Hayden and Check Hayden, the changes to the media ecosystem present new opportunities for science journalists to find work outside traditional news outlets. These include new digital popular science magazines, increasing demand for long-form writing and short-form videos and also new organisations that are developing innovative approaches to journalism. Examples include *The Conversation*, the online website whose editors work with academics to enable them to communicate directly with public audiences (Guenther and Joubert, 2021); the non-profit news website SciDevNet that supports science journalism in development contexts around the world; and intermediary organisations that sit at the interface between science and journalism, such as the Science Media Center Germany (see Chapter 5). Such initiatives benefit from financial support from scientific organisations, governments and charities that is not available to commercial news organisations. As the financial model of traditional journalism fractures, the historically porous boundary between science journalism and the institutions of science allows a re-positioning of the specialism in new ventures that are, in some ways, a reimagining of the initiative that helped establish science journalism in the United States a century ago (see Chapter 1).

Other types of intermediary organisation have long histories of their own. It would be a mistake to characterise scientific journals solely as part of the scientific community, publishing technical papers to serve that community and press releases to promote the science. Several prominent journals also have news divisions, staffed by science journalists. Thanks to their different business models, these can have the capacity to support in-depth journalism that is being squeezed at general news outlets. For instance, the *British Medical Journal* includes an Investigations Unit that supports investigative journalism into research misconduct, clinical ethics and other public interest issues.

The professional structures that have developed across a century of the science journalism beat also continue to offer support to science journalists. The World Federation of Science Journalists boasts over ten thousand members; many countries host awards for the best science journalism; and degree courses offer specialist training in science journalism. Many of the graduates of these courses might not find work on news desks, but as climate change, Covid and artificial intelligence (AI) all show, the need for specialists who can assess the reliability of science-related claims is greater than ever.

The crisis narrative emphasises the differences between science and journalism; the context of the post-truth era highlights that both are 'truth professions'. At their best, science and journalism share an aim to present accurate and evidenced accounts about the real world. Science journalists, given sufficient time and resources, have the expertise not just to simplify science, which implies it is reduced to something less than it is in technical discourse, but to add meaning by contextualising and analysing. The best science journalism contributes to a fuller understanding of science than scientists alone would produce.

The contents of the book

This book is divided into three parts. The first part examines the establishment of the beat during the twentieth century. Many of the issues that animate discussion about science journalism today have been present throughout its history. In Chapter 1, Susan Swanberg and I explore the emergence of the beat in 1920s America, where an innovative new organisation, Science Service, provided a news agency focused on science. By supporting specialist writers to make a living out of science news and producing science journalism that appeared in newspapers and on radio across the country, Science Service helped establish a demand for specialist science journalism. A few years after Science Service began its operations, *The New York Times* recognised science journalism's commercial viability by appointing its first staff journalist specialising in science. Both initiatives sought to encourage a scientific mindset and promote the importance of science to social progress. This alignment with the interests of the scientific community meant a blindness to the ideological work that science can do and led, particularly in the case of Science Service, to the promotion of oppressive eugenic ideas.

Like the US press, British newspapers also appointed science specialists in the period between the two world wars. However, it took longer for a journalistic approach to science to become established on radio in the UK, despite science being included in the broadcasting schedules from early on. As Jared Keller explains in Chapter 2, the BBC's remit to inform and educate meant that as late as the 1950s, the national broadcaster had a preference for scientists to give talks about their subject, rather than for journalists to interpret it. Keller takes four members of BBC staff, whose careers span the quarter of a century following World War II, to illustrate how the broadcaster moved slowly towards a more questioning and contextualising – a more journalistic – approach to science programming. The chapter reminds us not to take for granted that science should be subjected to a journalistic gaze.

Chapters 1 and 2 show how the first two generations of science journalists had to negotiate their roles not only with editors but also with scientists. Chapter 3 considers how they organised with each other to consolidate their specialism. Jane Gregory examines the role of professional organisations, mapping their emergence across continents and over time, from the establishment of the National Association of Science Writers in 1934 through to the digital networks of recent years. She shows how science journalists' formal associations have helped define the beat by providing their members with support through professional resources and a collective voice to champion both journalistic values and the importance of public engagement with science. Associations thrive, Gregory argues, when they draw on, but can be critical of, a strong news industry and a strong science base, enabling them to protect a commitment to the public interest in the face of conflicting corporate, political and economic pressures.

The second part of the book takes up this theme of how science journalists can safeguard their professional values by examining journalist-source relations in current science journalism. As discussed above, the dependency on PR materials is a central concern here. In Chapter 4, Lars Guenther, Justin Schröder and Anna Tratter examine 'churnalism', the practice of copying and pasting press releases into news stories with little or no alteration. They illustrate the high prevalence of the phenomenon in science journalism by tracking the take-up of press releases from the German Chemical Society. They suggest that a number of factors make science journalism particularly vulnerable to the influence of PR in the context of changing business models for news journalism. These include the growing investment in PR by scientific organisations, science journalists' reliance on expert sources and the beat's marginal role within newsrooms. Churnalism allows journalists to streamline their efforts in a pressured working environment, yet it also tips the balance of power towards the sources of news, threatening journalistic independence. In the current media environment, science PR meets the needs of both scientific organisations and news outlets, but in doing so, the public interest is de-prioritised.

In some countries, the relationship between science journalists and scientists is further mediated through intermediary organisations, known as Science Media Centres (SMCs). The first SMC was established in the United Kingdom in 2002 to serve as a press relations service for all of science; it immediately faced criticisms for supporting science PR at the expense of the public interest. SMCs in other countries have tried to address these concerns.

In Chapter 5, Irene Broer takes the case of SMC Germany, exploring how the organisation attempts to position itself on the journalism side of the journalism/science boundary. In its eight years of operation, SMC Germany has aimed to offer a service to news outlets by providing copy-ready science stories. In this, there are clear resonances with the aims of Science Service almost a century earlier. Broer presents ethnographic observations of the daily routines at SMC Germany in order to explore how the organisation navigates a path between journalism and PR.

Science PR is produced by scientific institutions and research-based companies. Intermediary organisations such as SMC Germany are similarly supported by universities, philanthropic organisations and industry. Both types of activities are predicated on a strong science base. In countries that lack significant levels of research activity, a further level of dependency is introduced as news outlets frequently turn to foreign sources to provide science news. In Chapter 6, An Nguyen and Minh Tran take Vietnam as an example to show how this over-reliance on foreign sources of news is manifested. Drawing on data from a sample of online science news and in-depth interviews with Vietnamese science journalists, they argue that the one-way flow of news from the Global North to the Global South is a form of neo-colonial knowledge dependency which, while helping raise awareness of global science developments, ignores the local impacts of science news and reinforces historical inequalities between the 'information haves' of the Global North and the 'information have-nots' of the Global South.

In authoritarian regimes, additional factors come into play as journalists of all types must operate under greater restrictions than outlets in liberal regimes. In Chapter 7, Alexandra Borissova Saleh considers the case of Russia and examines how science journalism operates when the media environment is dominated by state-owned outlets that are inclined towards a cheerleader role. Russia has a long history of science journalism, and in recent years, digital science magazines have offered a flexible alternative to state-owned media. However, in what Borissova Saleh describes as a 'one-two punch', first the Covid-19 pandemic and then the war in Ukraine created crisis conditions that made it both difficult and dangerous for Russian journalists to talk to news sources. The pandemic suddenly transformed science journalism from a soft news beat to one that dealt with highly politicised events; the war imposed even tighter controls on access to sources. Borissova Saleh argues that if, as in the case of Russia, scientists are unable to speak freely, then public expertise has died. Faced with this situation, Russian science journalists have been forced to choose between serving as state propagandists or risking their lives as activists.

Science journalists' dealings with the sources of science news are an important component of their work. The other is transforming the information so gathered into compelling and engaging output – a task that is all the more challenging given the technical and abstract nature of science. Part three of the book examines how this is accomplished by considering the voice of science journalism. News journalism in many countries strives for an objective and impartial voice. This is often achieved through the use of balance – the practice of quoting opposing sides in a matter of public controversy. In Chapter 8, I consider the meta-discourse about balance in science journalism, particularly as it has played out in the United Kingdom through criticisms of the use of 'false balance' in the BBC's coverage of climate change. I argue that the tendency of scientists and journalists to talk past each other is exacerbated by an underlying confusion between objectivity and impartiality. This leads scientists to disregard the need for broadcasters to represent public opinion and leads journalists to disregard the risks of contributors making false claims. The problem with balance in science journalism is not its overuse, as the criticisms of false balance imply, but rather its underuse and its perfunctory application when it is used. This has left science journalism vulnerable to manipulation when science becomes enrolled in ideological conflicts.

Chapter 9 examines how science journalists craft engaging stories from scientific developments. This often entails recasting scientific results as a journey of discovery told from a perspective that the audience can identify with. Katarzyna Molek-Kozakowska draws on trending articles from the UK's *New Scientist* magazine to illustrate the stylistic patterns and textual strategies through which journalists generate emotional engagement and newsworthiness. Molek-Kozakowska argues that journalists accomplish three distinct stylistic shifts as they adapt science from academic discourse to the popular science genre: an attentional turn, an affective turn and a narrative turn. These compositional devices and rhetorical repertoires construct a journalistic voice that must balance style and substance if it is to succeed in attracting audiences.

As Molek-Kozakowska shows, news journalists take a narrative turn, but a narrative approach is even more fully developed in long-form journalism. In Chapter 10, Lauren Kilian discusses the implications of narrative journalism when presented in the longest form – in non-fiction books. Literary science journalism often forgoes the objective voice of news journalism in favour of a subjective voice. Through the case study of Rebecca Skloot's prize-winning book *The Immortal Life of Henrietta Lacks*, Kilian explores the ethical compromises entailed in the narrative approach. In attempting to recount an immersive, multi-layered narrative, Skloot found that she had herself become part of the story. To be truthful about this, she needed to insert herself into the book's narrative; yet, Kilian argues, this authorial transparency had the unavoidable, and ethically damaging, countereffect of limiting the perspective of the narrative. Kilian concludes that the capacity of narrative journalism to deliver a level of immersion that is difficult to achieve in expository science journalism inevitably prioritises the demands of narrative over ethical considerations.

The book ends with a short conclusion in which I consider the future of science journalism through a conversation with one potential shaper of that future: an AI programme.

References

Abu El Rub, O., 2022. Forget superheroes – science journalists will save the world. *Al Jazeera Journalism Review*, 16 Jan., https://institute.aljazeera.net/en/ajr/article/1742.

Anderson, J. and Dudo, A., 2023. A view from the trenches: Interviews with journalists about reporting science news. *Science Communication*, 45(1), pp. 39–64.

Bauer, M.W. and Gregory, J., 2008. From journalism to corporate communication in post-war Britain. In: Bauer, M.W. and Bucchi, M. (eds.), *Journalism, Science and Society: Science Communication between News and Public Relations*. London: Routledge, pp. 33–51.

Bodmer, W., 1985. *The Public Understanding of Science*. London: Royal Society.

Brainard, C., 2008. CNN cuts entire science, tech team. *Columbia Journalism Review*, 4 Dec., https://archives.cjr.org/the_observatory/cnn_cuts_entire_science_tech_t.php.

Brumfiel, G., 2009. Science journalism: Supplanting the old media? *Nature*, 458(7236), pp. 274–277.

Buss, 2014. *Willy Ley, the science writers, and the popular reenchantment of science*. PhD thesis, University of Oklahoma.

Dunwoody, S., 2021. Science journalism: Prospects in the digital age. In Bucchi, M. and Trench, B., (eds.), *Routledge Handbook of Public Communication of Science and Technology* (2nd ed.). London: Routledge, pp. 14–32.

Fahy, D., 2017. Objectivity as trained judgment: How environmental reporters pioneered journalism for a "post-truth" era. *Environmental Communication*, 12(7), pp. 855–861.

Goldacre, B., 2005. Don't dumb down. *Guardian*, 8 Sep., Life section p. 4.

Guenther, L. and Joubert, M., 2021. Novel interfaces in science communication: Comparing journalistic and social media uptake of articles published by *The Conversation Africa*. *Public Understanding of Science*, 30(8), pp. 1041–1057.

Hansen, A., 1994. Journalistic practices and science reporting in the British press. *Public Understanding of Science*, 3(2), pp. 111–134.

Hayden, T. and Check Hayden, E., 2018. Science journalism's unlikely golden age. *Frontiers in Communication*, 2(24), 10.3389/fcomm.2017.00024.

Massarani, L., Entradas, M., Neves, L.F.F. and Bauer, M.W., 2021. *Global Science Journalism Report*. SciDevNet. https://www.scidev.net/wp-content/uploads/Global-Science-Journalism-Report-2021.pdf.

McKinnon, M., Howes, J., Leach, A. and Prokop, N., 2018. Perils and positives of science journalism in Australia. *Public Understanding of Science*, 27(5), pp. 562–577.

NASW, 2016. Final report of the NASW constitutional review ad hoc committee. NASW, 16 May, http://undark.org/wp-content/uploads/2016/05/NASW-Committee-Report.pdf.

Nature, 1961. A triumph for science. *Nature*, 190, 11 Apr., pp. 291–293.

Nelkin, D., 1987. *Selling Science: How the Press Covers Science and Technology*. New York: W.H. Freeman.

RSF, 2020. *Round-up 2020: Journalists Detained, Held Hostage and Missing*. Reporters Without Borders. https://rsf.org/sites/default/files/rsfs_2020_round-up_0.pdf.

Scripps, E.W., 1919. The American Society for the Dissemination of Science. In: Gardner, G. (ed.), *History of the Scripps Concern*. Ohio University Libraries, E.W. Scripps Papers, MSS 117, Series 4, http://bit.ly/scripps_mss117, 5 Mar., pp. 302–307.

Soth, A., 2020. "Beating the bounds". *JStor Daily*, 7 May, https://daily.jstor.org/beating-the-bounds/.

UN, 2020. During this coronavirus pandemic, 'fake news' is putting lives at risk: UNESCO. *UN News*, 13 Apr., https://news.un.org/en/story/2020/04/1061592.

Vogler, D. and Schäfer, M.S., 2020. Growing influence of university PR on science news coverage? A longitudinal automated content analysis of university media releases and newspaper coverage in Switzerland, 2003–2017. *International Journal of Communication*, 14, pp. 3143–3164.

PART I
Establishing a beat

1

"MAKING DEMOCRACY SAFE"

The development of US science journalism in the 1920s

Susan E. Swanberg and Felicity Mellor

In April 1913, an audience gathered at Columbia University's new School of Journalism in New York to hear about the opportunities available for popular science writers. The speaker, Edwin Slosson, expressed his dismay that so few journalists recognised science as a good source of news stories. Too often, he said, they regarded science with "contempt and ridicule" (p. 914). It was not only a problem of poor attitude, Slosson claimed, but also of inability: "when it comes to science itself, the ordinary reporter is at sea. He [sic] does not know the language of the laboratory. He does not know what he sees. He cannot distinguish between new and old, between the significant and the trivial. His nose of news has lost the scent" (p. 915).

Slosson had already started trying to revive this scent the previous autumn when he had begun teaching a course in physical sciences to Columbia's first intake of journalism students. He was well placed to do this, having worked for a decade as a chemistry lecturer and researcher before becoming literary editor of a New York weekly magazine, the *Independent*. He would continue the course at Columbia until he left the *Independent* in 1920 to become the first editor of a new organisation, Science Service. In his 1913 talk, Slosson had argued for journalists to take on the role of an "intermediary who can understand equally the language of the laboratory and of the street". Seven years later, Science Service aimed to fill precisely this intermediary role.

Historians have argued that Science Service played a foundational role in the development of science news journalism in the United States (Bennet, 2013; Ehrhardt, 1993; Foust, 1995; LaFollette, 2008, 2023; Nelkin, 1987; Olson, 2013; Swanberg, 2019, 2020a, 2021; Tobey, 1971). However, Slosson and the founders of Science Service were not alone in wanting to see more science in the newspapers. Although few histories have examined the science

DOI: 10.4324/9781003326724-3

reporting of *The New York Times* in this formative period, by the time Science Service started its work science articles were already a regular feature of the paper (Simpson, 1987; Unus, 2018). If Slosson flicked through the *Times* before he gave his talk that day, he would have seen stories about plans to eradicate mosquitoes, the travels of a sea serpent, a child prodigy and, on the first anniversary of the Titanic disaster, a front-page story about a steamship that had been saved from sinking by its double-hull design. The *Times*'s reporting of the sinking of the Titanic the year before had helped secure the reputation of the paper's managing editor, Carr Van Anda, who had realised before any other journalist that the ship must be lost (NYT, 1945). Under Van Anda's leadership, the *Times* came to devote more space to science than did any of its competitors (Berger, 1951: p. 250).

This chapter examines the emergence of American science journalism as a specialised beat in the 1920s by considering the distinctive contributions of Science Service and *The New York Times*: the first driven by a vision of the importance of science to a healthy democracy; the second by a vision of a new scientific approach to journalism based on facts and objectivity. The 1920s was a key period both in the development of newsroom culture more generally and also in the nationalisation of scientific research (Mari, 2021; Tobey, 1971). In this context, initiatives to encourage more and better science journalism fell on fertile ground. In the years following World War I, Science Service provided print-ready science news to newspapers unable to support specialist coverage of their own. A decade after it launched, about a hundred newspapers subscribed to Science Service's syndicated news bulletin and its newsletter reached about 12,000 subscribers directly (Bennet, 2013: p. 90). In the same period, Van Anda's *New York Times* became one of the first US newspapers to have a full-time science staff (Ehrhardt, 1993: p. 183). As Science Service cultivated a market for science news by serving as an intermediary between scientists and news journalists, newspapers such as *The New York Times* showed that commercial news outlets could support specialist science reporters of their own. Behind both developments was the shared conviction that accurate science news should be put before the public and the confidence that serious science stories could sell.

In what follows, we explore how this was put into practice. The first specialist science journalists had to negotiate the competing demands of scientists who wanted to control how their field was reported and editors who wanted dramatic – perhaps even sensational – stories to entice readers. Slosson and his colleagues at Science Service, and Van Anda and his reporters at *The New York Times*, set about producing science news that was both accurate enough to satisfy scientists and also dramatic enough to attract newspaper readers. They were able to do so because of the changing social position of science brought about by the war. The compromise they reached – working closely with scientists by promoting the social benefits of science while also conforming to journalistic news values – established the prevailing approach of American science journalism for most of the next century. However, it was a compromise that entailed an implicit – and sometimes explicit – ideological positioning under the guise of the new 'objective' journalism. The emerging science beat would be shaped by the idea that science was not merely useful but that scientific ideas were the necessary source of social progress.

"A new agency for the popularization of science"

As American civil society regrouped in the years immediately following World War I, leading scientists became concerned about the public's attitude towards science. Historian Ronald Tobey (1971) has argued that in centralising the organisation of science and redirecting research towards the national agenda, the war persuaded scientists that they had a role to play in wider society regardless of how specialist their expertise might be. However, some feared that the destruction that the war had wrought – and the role that science and technology had played in this – had left the public wary of the impact of science. Leading American scientists responded to this situation by emphasising the relevance of the values of professional science to those of American democracy. As Tobey argues, this was an ideological move predicated on a vision of a progressive national culture.

Retired newspaper magnate E.W. Scripps shared scientists' anxiety about American democracy and their belief that science had a fundamental role to play in saving it. Scripps believed that the development of what philosopher John Dewey (1910: p. 127) had called a "scientific habit of mind" was necessary for democracy to thrive. Scripps perceived "a vast quantity of misinformation being constantly spread abroad" by the nation's newspapers (1919: p. 306). As the owner of the first newspaper chain in the United States and one of the most influential newspaper men of his time, Scripps understood the power of the press to inform (and misinform) the public, and he wanted newspapers to do more to cultivate a scientific way of thinking in their readers. To this end, in 1919, he began to plan a new organisation to do just this.

Scripps turned for help to his friend and associate William Ritter, the scientific director of the Scripps Institution for Biological Research, which Scripps and his sister Ellen had financed when it was established in 1903. The new organisation, originally called the American Society for the Dissemination of Science, was to present scientific findings in a form accessible to the sorts of people to whom Scripps's newspapers appealed: in Scripps's words, "the 95%, namely the working man and the poor and unfortunate" (Bennet, 2013: p. 48). In 1902, he had founded the Newspaper Enterprise Association, a syndicated news service that supplied his own newspapers – and soon hundreds of other papers – with content that they could not afford to produce themselves.

His vision for the new organisation was similar, but with a focus on science content, and he anticipated that the Newspaper Enterprise Association would be one sure customer (Scripps, 1920: p. 293). It would not be easy, however. Scripps noted that science was "something that almost the universality of ordinary men on the street believe to be not only uninteresting, but not at all worth while." The new Society, he admitted, would have to be "patient and perhaps a little crafty in stealing its way into public regard and confidence" (p. 295).

Scripps provided the initial funding for the organisation, donating $30,000 a year with the promise of a further half million dollar endowment if all went well. It was important to Scripps that, to ensure its independence from commercial interests, the organisation be non-profit, but, once established, self-financing through the services it sold (Scripps, 1919: p. 303). It was also important to him that the organisation gained the trust of scientists in order to encourage scientists and journalists to overcome their suspicions of each other (Bennet, 2013: p. 55).

The new organisation, now renamed Science Service, started work in January 1921, with offices at the National Research Council, a clear indication that the venture had the support of the scientific establishment. Indeed, the Board of Trustees consisted primarily of scientists – nine in total, drawn from three national science organisations, with zoologist Ritter appointed as the Board's first president. In a change to Scripps's original plan for a board composed exclusively of "men of science", the trustees also included three journalists and three representatives of the Scripps estate (Scripps, 1919: p. 302; Slosson, 1921: p. 322).

The Board's first task was to appoint an editor. As both a scientist and a journalist, Edwin Slosson was an obvious candidate to play a key role in bridging science and journalism, especially given his experience teaching science to journalists at Columbia University and the recent success of his popular book *Creative Chemistry* (1919). Scripps was not entirely satisfied with Slosson's style of writing – thinking that it could be too dry and inaccessible – and other candidates were considered, but after some negotiations, Slosson was appointed. He was to run the service until his death in 1929 (Ehrhardt, 1993; Foust, 1995).

In a talk to the Board of Trustees in 1921, Slosson explained that his plan was "to crowd out falsehood by truth and to present scientific information in a way that will be at least as attractive as the misinformation which now holds the field" (Bennet, 2013: p. 84). He did this by producing the *Science News Bulletin*, a weekly (and later daily) collection of syndicated science news reports edited by Watson Davis, an engineer and columnist who Slosson had recruited as his principal writer. In mailings advertising a four-week trial to newspaper editors, the service offered itself as the "reporter" on a newspaper's science "beat" (Bennet, 2013: p. 78). Sixteen newspapers subscribed to the offer, entitling them to publish the *Bulletin* reports either verbatim or with additions or subtractions.

The first edition of the *Bulletin* in April 1921 carried a front-page story about the Prince of Monaco receiving the Agassiz medal for his scientific researches. This was followed by 11 shorter articles covering a wide range of subjects, including a humorous report on America's stalest bread (a 500-year-old tortilla from Colorado), the discovery of a 20-mile-long fossil bed, and the loss of wildlife (and consequent economic losses) resulting from the grass fires set each spring to clear fields. Illustrating the accessible style that Science Service was aiming for, a chirpy piece offering advice from "Uncle Sam's experts" on how to make housework less of a chore ended with the optimistic suggestion to: "Make the family help" (Science Service, 1921a).

Within six months of its launch, 30 newspapers across the United States had signed up to receive the *Bulletin*, giving Science Service a potential reach of some 1.5 million readers (LaFollette, 2008: p. 61). Slosson did not originally intend for the Service to produce a periodical of its own. However, in response to requests from the public, in July 1922, Science Service launched a newsletter that was mailed directly to subscribing individuals, schools and libraries. The *News-Letter* contained some of the same news as the *Bulletin*, although frequently in expanded form, and also introduced some new features. Within a few years, the *News-Letter* had attracted around a thousand subscribers. Eventually, it would become the organisation's primary printed vehicle for science news and it continues to be published today under the title *Science News*. Science Service was quick to see the importance of radio, not only covering technological developments in the *News-Letter* but also producing radio broadcasts, scripts and recordings, starting in 1924 with a series of talks by distinguished scientists (LaFollette, 2008). As Science Service became established, it also moved into book publishing and educational projects to engage young science students through science kits, science clubs and science fairs (Bennet, 2013; Davis, 1948).

Before joining Science Service, Slosson's time in New York had brought him into contact with a new circle of intellectuals, including John Dewey, whose participatory model of democracy would inform Slosson's vision for Science Service (Tobey, 1971: p. 72). Slosson, Scripps and Ritter all shared the belief that science had a fundamental role to play in safeguarding democracy. As Slosson explained when he announced the launch of Science Service in the journal *Science*:

> The success of democratic government as well as the prosperity of the individual may be said to depend upon the ability of the people to distinguish between real science and fake, between genuine expert and the pretender.
> (Slosson, 1921: p. 321)

Watson Davis, who led the Service after Slosson's death in 1929, agreed: "So many of the ideals that we cherish, such as liberty, opportunity, the pursuit of happiness, freedom, democracy, are achieved by the utilization of scientific methods.... The mistakes of science and democracy are best corrected by the methods of science and democracy" (Davis, 1948: p. 241). Scripps, too, had been motivated by the belief that science was essential to a healthy democracy. As Ritter recalled on the death of his friend in 1926, when they had first discussed the idea of a science news agency, Scripps had justified it on the grounds that: "the only way of making democracy safe is to make it more scientific" (Ritter, 1926: pp. 7-8). Indeed, Ritter saw in Science Service an expression of Scripps's overarching philosophy – a philosophy that Ritter characterised as "perennially buoyant objective realism" (Ritter, 1926: pp. 7-8).

Science journalism at *The New York Times*

Scripps's approach to newsgathering was also anchored in that same worldview: "Get the truth just as far as possible; write the essence of it so simply and clearly that anybody who can read it at all can understand it; and publish it so cheaply that anybody who has any money at all can afford to buy the narration" (Ritter, 1926: p. 12). The Scripps newspaper empire therefore targeted ordinary workers. At *The New York Times,* by contrast, a similar aspiration to report the truth was attracting a more elite audience under the famous slogan: "All the news that's fit to print". The failing paper had been acquired in 1896 by Adolph Ochs, owner of the *Chattanooga Times* in Tennessee. Ochs immediately announced his desire to "conduct a high-standard newspaper, clean, dignified and trustworthy", as he set about reviving the paper's circulation by distinguishing it from the sensationalist 'yellow journalism' of its competitors (Schudson, 1978: p. 110).

By 1920, Ochs had increased the paper's circulation from 9,000 to 320,000 and had secured its reputation as a bastion of journalistic integrity (Porwancher, 2011). In this, he was helped by his managing editor, Carr Van Anda. According to journalist Meyer Berger's history celebrating the paper's centenary, Ochs and Van Anda shared not only a desire for unsensationalised journalism but also "the same eager curiosity about the unknowns in the sciences and about the remote unexplored corners of the world" (1951: p. 160). The two men agreed that 'all the news that's fit to print' needed to include science news.

Van Anda had enrolled at Ohio University at the age of 16 to study mathematics and physics, but had left early to pursue his interest in newspapers. Later in life, Van Anda's mathematical abilities became legendary. In a biography marking Van Anda's retirement, Barnett Fine described him as "a human volcano, a mortal encyclopedic fury of uncanny intellectual ability" (1933: p. 19). Fine recorded how Van Anda had astonished colleagues on one occasion by translating some Egyptian hieroglyphs when no Egyptologists had been available. On another occasion, the British Museum accepted his correction of a translation by one of its own experts. Van Anda also corrected

Albert Einstein, spotting a mistake in an article based on a lecture by him. Asked for confirmation, Einstein agreed he had made an error when copying out an equation. As this suggests, Van Anda was capable of engaging scientists on their own terms, as when the journal *Science* published a lengthy technical article by him criticising the theories of the origin of the solar system developed by two eminent British physicists (Van Anda, 1931).

As Ochs's managing editor, Van Anda had the freedom to cultivate his scientific interests within the newspaper. He ensured the *Times* reported on technological advances, such as human flight, that were being ignored by other papers and he developed a keen 'nose of news', as Slosson would have put it, that could sniff out news values in even the most abstract of subjects. Notable in this regard was his recognition that the solar eclipse measurements testing Einstein's theory of relativity were a major story. The paper ran one article mentioning the possible proof of Einstein's theory at the time of the eclipse in 1918, and a long series of articles after the announcement of the findings in London the next year (Campbell, 1918; NYT, 1919). Although *The Times* of London reported the 1919 announcement first and devoted more coverage in the months running up to the event (Sponsel, 2002), *The New York Times*'s saturation coverage afterwards helped transform Einstein into an international celebrity.

Thus, by the time Science Service was launched in 1921, *The New York Times* had already firmly established its interest in science news. The following year, Van Anda assigned one of his general news reporters, Alva Johnston, to cover the three-day annual meeting of the American Association for the Advancement of Science. Newspapers had carried news of scientific meetings since the previous century, but the reports had usually come from scientists attending the meetings. This year, the Association's meeting was attended not only by journalist Alva Johnston, but also by Edwin Slosson and Watson Davis from Science Service and David Dietz from the Scripps newspapers. It was, however, Johnston's reporting that stood out. Realising that Van Anda would give him all the space he could fill, Johnston filed detailed accounts of the talks, including what was to become the paper's first article explaining atomic energy. Johnston was awarded the 1923 Pulitzer Prize for his coverage of the meeting (Unus, 2018: p. 12).

Despite such successes and having sufficient resources to investigate science stories itself, *The New York Times* also occasionally carried Science Service articles, such as a 1932 story about a professor who had found that an 800-year-old wine was unfit to drink (Science Service, 1932). By the time of that story, however, the *Times* could boast its own specialist science journalists. The Scripps newspapers had already taken this step in 1921 when David Dietz, a reporter who had covered science at Scripps's *Cleveland Press* since 1915, became science editor for the whole Scripps-Howard chain, likely making him the first full-time science editor at a US newspaper (Dietz, 1966: p. 13; Ehrhardt, 1993: p. 176). The *Times* finally committed to a dedicated science specialist six years later, in 1927, when Waldemar Kaempffert became the paper's first science editor (Davis, 1957). Kaempffert left a year later to set up the Chicago Museum of Science. Despite most newspapers avoiding hiring specialists of any type at this time due to the deepening financial crisis, a couple of years later William L. Laurence was appointed to cover the science beat. Laurence later recalled being told that science was "the one spot we're weak in.... We need a specialist. We need someone to concentrate on science" (Ehrhardt, 1993: p. 185). According to Meyer Berger, prone to overstating the *Times*'s achievements, Laurence was the "first newspaper reporter assigned exclusively to daily science coverage" (1951: p. 276); Kaempffert and Dietz having had editorial roles. Certainly, when Kaempffert returned to the *Times* in 1931, the paper's retention of two science specialists made its commitment to science reporting clear.

As well as positioning science as a newsworthy subject, Van Anda also brought what was perceived as a scientific approach to the process of newsgathering. According to one journalist who worked under him, Van Anda regarded newsgathering as "an intellectual problem": he would map out the problem and track down the news as if doing "a piece of mental gymnastics" (Unus, 2018: p. 8). Van Anda's scoop on the Titanic disaster showed how this sort of problem-solving, based on small scraps of evidence, could deliver a story before competitors. His personal interests in Egyptology similarly enabled him to spot the newsworthiness of the excavation of Tutankhamun's tomb, ensuring the paper secured exclusive US rights to what would become a long-running story (Fine, 1933: p. 72).

Noting the lack of adequate biographies over a decade after his death, historian of journalism Allan Nevins suggested that Van Anda's approach to news work was inspired by science:

> He saw how complex the truth is, and realized that to discover it a great newspaper must have not simply a slick skill in reporting surface news, but a patient, scientific-minded exploration, by well educated specialists, of intricate situations. An event is a force momentarily made visible. The good news specialist must look for the force behind the event, as something to be explored, measured and analyzed.
>
> (1959: pp. 416-417)

In his short volume commemorating Van Anda's retirement in 1933, Benjamin Fine, later himself a journalist at the *Times*, drew a similar conclusion based on the interviews he had conducted:

> Getting news to Mr. Van Anda was a scientific undertaking, to be achieved by intellectual conquest. He analyzed, dissected, burrowed, into the story under consideration, and then knew precisely what steps should be

followed in order to get the best results. ... He believed that managing a paper could be done as scientifically and as intelligently as running a laboratory.

(Fine, 1933: p. 44)

The new science journalists

To fulfil their ambitions for increased and improved science coverage, both Edwin Slosson at Science Service and Carr Van Anda at *The New York Times* needed writers capable of covering complex subjects accurately. Before World War I, science writers had generally also been scientists: Slosson himself typified this combination. And so it was to scientists that Slosson initially turned for articles for Science Service. Indeed, his desire to find writers who were scientists led him to hesitate over hiring Watson Davis as his assistant. Davis wrote a regular science column for the *Washington Herald* and also worked as an assistant engineer at the National Bureau of Standards, yet Slosson worried that this was not close enough to cutting-edge research. As a result, for the first year and a half, Davis was hired on a part-time basis while continuing to work full-time at the Bureau. After Slosson died in 1929, some of the scientists on the Board of Trustees again hesitated over Davis's lack of research experience. It would take another four years before he was formally promoted to the position of director (Bennet, 2013: p. 67, 91).

Slosson included an appeal for writers when he announced the launch of Science Service in the pages of *Science*, informing the journal's scientist readers that he hoped to secure "correspondents in every university and center of research who have the time, disposition and ability to write for non-technical journals." He was particularly keen, he said, to be in touch with "young men and women in the various sciences who have literary inclinations and would be willing to submit to a rigorous course of training with a view to making the writing of popular science a part of their life work" (Slosson, 1921: p. 323). Slosson made no mention of the pay they might expect to receive, but Scripps's original proposal had estimated that contributors would be paid \$20 per thousand words (Scripps, 1919: p. 304). If a small number of contributors did all the writing, Scripps had calculated, they would earn between \$500 and \$1,000 a year. At a time when the average reporter was earning over \$2,000 a year (Mari, 2021: p. 240), this confirms that the work was to be treated as a supplement to other employment.

However, the calls for active scientists to come forward attracted little interest and those who did come forward could not write well enough. After a year of scouring the United States for science writers, Slosson complained that he had had to turn to England to find them (1922: p. 241). Without scientists providing material, Slosson had to rely on journalists. He ensured this did not lead to inaccuracies by also recruiting staff writers to maintain quality. For these staff writers, he still looked to those with a scientific education. Of the four writers on the staff by the end of 1926, three had been trained in science. These writers' work would become the most widely distributed science news in the country.

The New York Times seems not to have shared the assumption that a scientific education was necessary for a journalist to cover science. Pulitzer Prize-winning Alva Johnston was a general reporter with no previous experience in science. Waldemar Kaempffert did have a bachelor's degree in science, had trained as a patent lawyer, and had been assistant editor of *Scientific American* and editor of *Popular Science Monthly* before he joined *The New York Times* (Davis, 1957). However, William Laurence had studied law before covering aviation stories at the *New York World*; he got the job of science reporter at the *Times* on the basis of one story about a critique of Einstein's theory (Kiernan, 2022: pp. 3-4). Both men went on to have long and celebrated careers as science journalists. Laurence was to win two Pulitzer prizes, co-found the National Association of Science Writers (for which both Laurence and Kaempffert served as presidents; see Chapter 3) and was the only journalist to witness the dropping of the atomic bomb on Nagasaki while seconded to the Manhattan Project (NYT, 1977). Similarly, when Kaempffert died, still working as the *Times*'s science editor at the age of 79, the paper emphasised his contribution to the professionalisation of science journalism: "If he did not create the profession of science writer, he certainly, over the last half century, invested it with new standards of ethics, scholarship, dignity and usefulness" (NYT, 1956).

It is noteworthy that when Slosson put out his call for science writers, he had appealed to both "men and women" to come forward (Slosson, 1921: p. 323). By this time, women were starting to make in-roads into newspaper journalism. US government labour statistics for 1920 show 5,273 women working as reporters and editors, about a quarter of the total newsroom workforce (Steiner, 1997: p. 4). At the start of the century, women had been largely confined to reporting on women's issues or, more daring but much denigrated, were employed as "stunt girls" producing exposés based on what we would now call undercover reporting. As journalist Ishbel Ross put it in her early history of women journalists: "Newspaper women had to go in for dizzy self-exploitation before they could make themselves heard at all" (1936: p. 14).

There was little scope for stunts in science reporting, but science writing did offer an opportunity for women otherwise excluded from science careers to make a living by communicating science. One such woman was Mary Proctor, who, after following her father's career as a popular lecturer in astronomy, became a regular contributor to *The New York Times* shortly after Van Anda took over as editor. Between 1905 and 1909, Proctor wrote 24 columns for the *Times* and in 1910, she became the paper's special correspondent reporting on Halley's Comet. Her reports for the *Times*, detailing the latest research based on interviews with the researchers, were republished across the country. Historian Martin Bush identifies Proctor as a pioneer of a form of reporting – "point-of-view reportage based on personal connections with researchers" – that would become a staple of science journalism (Bush, 2023: p. 72).

Had she applied, however, Proctor would have been unable to get a job as a staff reporter at the newspaper that published her columns, for Ochs refused to hire women reporters. Throughout Ochs' tenure at *The New York Times*, only three women worked as staff reporters at the city desk, with Van Anda's night secretary also occasionally contributing news reports (Ross, 1936: p. 150). Women journalists – like Mary Proctor or foreign news correspondent Anne O'Hare McCormick, who was finally hired on the staff after Ochs died – were able to contribute as freelances only.

Slosson had a different view of the contribution that women could make. In contrast to Ochs, Slosson supported women's rights. His wife, May Preston Slosson, had been the first woman in the United States to gain a doctoral degree in philosophy and was an active suffragist (Wilensky, 2021). That Slosson approved of his wife's campaigning is evidenced by his taking her place when she was too ill to give a talk at a suffrage meeting in New York. He assured the gathering that giving women the vote, as had already happened in some other states, would neither harm women nor "destroy the home" (NYT, 1910). In a later *Science News-Letter* article about biological and cultural inheritance, entitled 'How women control the future', Slosson noted in passing that political enfranchisement was "an act of justice and will doubtless prove beneficial" (1924a: p. 6).

From the start, Slosson envisaged a role for women at Science Service, albeit in terms that seemed to merge editorial work with clerical functions. He proposed that the new organisation should employ assistants who were "bright and careful women familiar with scientific journals and the terminology of the several sciences". They would need to be capable of researching topics in various languages, editing copy and writing "snappy captions", as well as doing the typing and mimeographing (Slosson, n.d.: p. 309). Once Science Service began operations, Slosson recruited women onto the writing staff, as historian Marcel LaFollette (2023) details in her comprehensive history of these journalists. Early recruits included Emma Reh, a chemistry graduate; Emily Cleveland Davis, who specialised in archaeology but wrote about other topics as well; and Marjorie MacDill, who had majored in biology and English and wrote about ecology and zoology.

Most well known for her contribution to the emerging beat was Jane Stafford, who joined Science Service in 1928 as its medical sciences specialist when MacDill left. Stafford had majored in chemistry, worked as a laboratory technician in a hospital and as an assistant editor at the American Medical

Association's popular journal *Hygeia* before joining Science Service. She became one of the co-founders of the National Association of Science Writers (despite initial scepticism about the group's purpose) and served a term as its president as well as serving as president of the Women's National Press Club (LaFollette, 2023).

The last writer hired during Slosson's editorship was Marjorie Van de Water, who had worked at the National Bureau of Standards and had developed psychological tests at the National Research Council and the US Civil Service Commission before becoming Science Service's psychology writer. Van de Water won the American Psychological Foundation's 1959 award for science writing and stayed with Science Service until her death in 1962 (Swanberg, 2020a).

Watson Davis continued the policy of supporting women after Slosson's death. Indeed, when Davis had first tried to persuade Slosson to hire him, he had offered his wife's services as well, since she had an undergraduate degree in chemistry (Bennet, 2013: p. 67). One of the first series of syndicated columns that the Service offered, about the history of science, was written by Helen Miles Davis and she would become editor of a later Science Service off-shoot *Chemistry Magazine* (Ehrhardt, 1993: p. 155, 310).

By prioritising a scientific background in its staff writers, Science Service was able to enjoy the support of the American institutions of science. As these appointments show, this requirement did not result in a bias towards male writers, despite – indeed, perhaps because of – the continued male dominance of science. On the contrary, as women were now able to study scientific subjects at university but still faced barriers to research careers, the emerging field of science journalism offered them opportunities to pursue their interests. In direct contrast to *The New York Times*, Science Service provided a safe haven for women to develop careers as science writers. These women made up half the Service's workforce throughout the 1920s and were integral to the development of science journalism as a professional specialism.

"Not a propaganda"

If, from our current vantage point, Science Service fell on the right side of history in its support for women science journalists, on another important issue – eugenics – the organisation got it badly wrong. In his original plan for the organisation, Scripps had been clear: "The first aim of this institution should be just the reverse of what is called propaganda" (Scripps, 1919: p. 306). The organisation would, he proposed, present facts on the basis of which people could form their own opinions. Slosson agreed, emphasising that policy debates needed to be grounded in "accurate data". Timely publication of such information could have "a strong influence on public policy", he noted, "yet if the facts were presented in a fair and unbiased manner no one could criticise the Service as partisan or meddlesome" (Slosson, n.d.: pp. 314-315). This view of science as a producer of neutral facts was manifested in a reliance on single-source reporting and a failure to interrogate scientific claims, even when they were being deployed to promote ideological agendas. Scripps had promised that his organisation was to be "Not a propaganda" (Scripps, 1919: p. 306), yet from the start it served up eugenic propaganda calling for policies of forced sterilisation of disadvantaged groups.

This was not, however, simply the unintended outcome of journalistic methods that avoided asking probing questions. The Service's uncritical coverage of the eugenic movement over many years aligned with Scripps's own views of eugenics. Scripps's unpublished private papers reveal his support for the movement, including some of the most extreme measures to prevent the disadvantaged from bearing children. Like other respectable progressives in the decades before Nazi Germany took such reasoning to its abhorrent endpoint, Scripps's support for the working class was accompanied by a belief that scientific breeding could solve social problems by improving the quality of the human race. Scripps's eugenic advocacy was mirrored at Science Service. While Ritter came to reject eugenics, despite some initial interest in hereditarian thinking, many trustees and staffers of Science Service were eugenic partisans (Olson, 2013: p. 296). For example, Watson Davis was actively recruited onto the board of the American Eugenics Society and served in that role for decades during his time at the Service (Swanberg, 2021).

This support for eugenics was reflected in the Service's output. One early newspaper article with the Science Service byline repeated a recommendation by the son of Charles Darwin that the "well-born" should raise large families to "arrest the racial deterioration that is taking place because of the multiplication of the inferior and ill endowed" (Science Service, 1921b). *Science News-Letter* also regularly published pro-eugenics articles throughout the 1920s. For instance, a typical 1924 article about immigration controls repeated statements in a report by the Eugenics Committee without comment. Had mental tests been used to control immigration, the article reported, over six million "aliens", now able to become parents of future Americans, would have been barred from entering the country. "It is high time", the article quoted, "for the American people to put a stop to such degradation of American citizenship" (Science Service, 1924). Later the same year, Edwin Slosson reported the views of an Oxford and Harvard psychologist that the decline of the nation's "racial stock" needed to be halted by paying men of "proved ability" to have more children. Slosson's front-page article noted disagreement about what measures would be most appropriate, but not that some disagreed with the reasons given to justify such measures (Slosson, 1924b).

The Service continued to cover the topic, in a largely uncritical manner, over many years. The output included a book chapter on 'Race Betterment' by Jane

Stafford, published at a time when many US states had introduced eugenic sterilisation laws and the Nazi Party had recently enacted compulsory sterilisation laws in Germany. In addition to setting out the supposed arguments in favour of eugenics, Stafford did also set out counterarguments, noting that opponents regarded eugenic sterilisation as "unscientific and unfair" (Stafford, 1934: p. 278). However, a short *News-Letter* article on the new German laws offered no such criticism or context, instead reporting the approval of US eugenicists who were confident that the laws could not "be made 'an instrument of tyranny' for the non-Nordic races" (Science Service, 1933). As late as 1939, Science Service organised a survey of scientists asking the question: "How could the world's population be improved most effectively genetically?" The survey provoked a critical response from leading biologists who rejected the genetic determinism implied by the question and who called for social reforms, rather than purely biological interventions, to create a more just society (Crew et al., 1939). Only as Europe descended into war did Science Service's coverage of eugenics finally decline. In 2022, its centenary year, *Science News-Letter* – now renamed *Science News* – issued an apology for its racist coverage of eugenics in the 1920s and 1930s (Science News Staff, 2022).

The New York Times was less consistent in its treatment of eugenics. Like other newspapers, it covered the topic extensively from 1912, often in enthusiastic terms, and gave a platform to eugenicists. A typical headline 'Scientists see eugenics aid in doing away with crime' gives some flavour of the coverage (NYT, 1923). Yet, as legislation was debated, the paper's commitment to political neutrality meant it also had to report opposition to the laws and its coverage included criticisms of prominent eugenicists (Cooke, 1998).

Before he had joined the *Times*, as editor of *Popular Science Monthly*, Waldemar Kaempffert had been enthusiastic about eugenics, writing to one prominent eugenicist in 1915 that he planned a series of articles "discussing the incalculable social harm done by bad human protoplasm" (Gillette, 2007: p. 133). However, Kaempffert's position had shifted by the time he was working at the *Times*. In 1928, he wrote a long feature for the paper's magazine that exposed some of the flawed notions underpinning eugenics (Kaempffert, 1928). In a book review of 1932, he complained that eugenicists "usually know nothing of genetics" and he endorsed those who argued against genetic determinism (Gillette, 2007: p. 133). In a letter to the editor of *Eugenics Review* a few years later, he clarified his view: "I am an opponent not of eugenics, but of the nonsense that is uttered in its name by far too many American social reformers" (Kaempffert, 1936).

Conclusion

Beat reporting began to emerge in the United States at the end of the nineteenth century as part of the professionalisation and rationalisation of the trade. However, it was only after World War I that science developed as a specialist beat when prominent scientists began to emphasise the role of science in social progress, ushering in what Watson Davis referred to as "an era of good feeling between scientist and newspapermen" (Davis, 1948: p. 240).

Science Service played a key role in this transformation, producing science news that was published throughout the US. The Service's success in placing copy helped to create a market for science news and to establish the form it would take: entertaining summaries of research findings. At the same time, elite newspapers such as *The New York Times* found that science news could serve as a means to distinguish themselves from more populist outlets, emphasising the fact-based nature of their reporting while simultaneously delivering dramatic stories that transported their readers to unknown places or thrilled them with new ideas. The editors of both organisations understood the importance of applying standard journalistic news values when covering science and believed that this was possible. As one advert for Science Service put it: "Drama lurks in every test tube" (Nelkin, 1987: p. 90).

Scientists seem to have been happy with the efforts of both organisations. At a 1933 symposium at the University of Wisconsin where scientists and journalists discussed the difficulties in their relationship, the chair of the physics department commended Science Service for doing "very valuable work and doing it well", while the chair of the zoology department singled out *The New York Times*: "I have no quarrel with journalists in science. The news service is improving steadily. Such papers as the New York Times are doing a very good job in reporting scientific meetings and other like items" (Quill, 1933). When, in 1945, in place of its usual annual award, the National Association for Science Writers commended 12 journalists for "career excellence in science reporting", Waldemar Kaempffert and William Laurence were both cited, as were four Science Service writers – including Jane Stafford and Marjorie Van De Water, the only women on the list (NASW, n.d.).

Science Service had been conceived as an intermediary organisation, bridging the worlds of science and journalism. Its founders believed the support of scientists was essential to the development of accurate and informative science reporting. As a consequence, they sought out science journalists who were trained in science and they shared with many prominent scientists a worldview that saw science as the most reliable instrument for social progress. This worldview may have informed Science Service's support for women seeking careers as science journalists, but it also led to a moral blindsight that enabled its uncritical support for eugenics.

The New York Times also promoted shared values with science, albeit in a subtly different way. There, science aligned with the paper's targeting of an elite audience through its appeal to objective journalism. The idea of 'news from nowhere' cohered with science's similar appeal to universal knowledge. This apparent compatibility was taken to an extreme during World War II when William Laurence acted as press agent for the Manhattan Project while still on the *Times* payroll (Swanberg, 2020b).

In his talk at the Columbia School of Journalism, Edwin Slosson summed up the motivation, as he saw it, behind the development of science journalism: "More than all else the journalist needs the spirit of science, the vital impetus of perpetual progress and the limitless vision of the future" (Slosson, 1913). In 1920s America, such aspirations enabled science journalism to become established as a specialist beat. Science specialists found their place in the newsroom, but only by working on a beat whose independence had been compromised by aligning science news with that "limitless vision of the future" offered by some scientists. It was a compromise that science journalists would have to continue to navigate for the next century.

References

Bennet, C., 2013. *Science Service and the Origins of Science Journalism, 1919–1950.* PhD thesis, Iowa State University.

Berger, M., 1951. *The Story of the New York Times, 1851–1951.* New York: Simon and Shuster.

Bush, M., 2023. Mary Proctor: An astronomical popularizer in the shadows. *Notes and Records*, 77(1), pp. 49–72.

Campbell, W.W., 1918. Clouds fall away for solar eclipse. *New York Times*, 10 Jun., p. 7.

Cooke, K.J., 1998. The limits of heredity: Nature and nurture in American eugenics before 1915. *Journal of the History of Biology*, 31(2), pp. 263–278.

Crew, F.A.E. et al., 1939. Social biology and population improvement. *Nature*, 144, pp. 521–522.

Davis, W., 1948. The rise of science understanding. *Science*, 108(2801), pp. 239–246.

Davis, W., 1957. Dr. Waldemar B. Kaempffert. *Nature*, 52(2), pp. 17–18.

Dewey, J., 1910. Science as subject matter and as method. *Science*, 31(787), pp. 121–127.

Dietz, D., 1966. Science, newspapers, and the future. *The Quill*, 54(7), pp. 12–17.

Ehrhardt, G.R., 1993. *Descendants of Prometheus: Popular Science writing in the United States, 1915–1948.* PhD thesis, Duke University.

Fine, B., 1933. *A Giant of the Press.* New York: Editor & Publisher Company.

Foust, J.C., 1995. E.W. Scripps and the science service. *Journalism History*, 21(2), pp. 58–64.

Gillette, A., 2007. *Eugenics and the Nature-Nurture Debate in the Twentieth Century.* New York: Palgrave Macmillan.

Kaempffert, W., 1928. The superman: eugenics sifted. *The New York Times*, 27 May, pp. 72–73, 83.

Kaempffert, W., 1936. Eugenics in USA. *Eugenics Review*, 28(2), p. 164.

Kiernan, V., 2022. *Atomic Bill: A Journalist's Dangerous Ambition in the Shadow of the Bomb.* Ithaca: Cornell University Press.

LaFollette, M.C., 2008. *Science on the Air: Popularizers and Personalities on Radio and Early Television.* Chicago: University of Chicago Press.

LaFollette, M.C., 2023. *Writing for Their Lives: America's Pioneering Female Science Journalists.* Cambridge: The MIT Press.

Mari, W., 2021. *The American Newsroom: A History, 1920-1960.* Columbia: University of Missouri Press.

NASW, n.d. Award winners: 1945. *AAAS Kavli Science Journalism Awards*, https://sjawards.aaas.org/awards/field_award_year/1945-13.
Nelkin, D., 1987. *Selling Science: How the Press Covers Science and Technology*. New York: W.H. Freeman.
Nevins, A., 1959. American journalism and its historical treatment. *Journalism Quarterly*, 36(4), pp. 411–422, 519.
NYT, 1910. Husband talks suffrage for her. *The New York Times*, 12 Jan., p. 9.
NYT, 1919. Eclipse showed gravity variation. *The New York Times*, 9 Nov., p.6.
NYT, 1923. Scientists see eugenics aid in doing away with crime. *New York Times*, 29 Jul., p. 3.
NYT, 1956. Waldemar B. Kaempffert dies; science editor of the Times, 79. *The New York Times*, 28 Nov., p. 35.
NYT, 1977. William Laurence, ex-science writer for the Times, dies. *The New York Times*, 19 Mar., p. 47.
Olson, A.I., 2013. *The People's Classroom: American Modernism and the Struggle for Democratic Education, 1860–1940*. PhD thesis, University of Michigan.
Porwancher, A., 2011. Objectivity's prophet. *Journalism History*, 36(4), pp. 186–195.
Quill, 1933. Additional scientists and newspapermen discuss difficulties in their dealings. *The Quill*, 21(3), pp. 8–9.
Ritter, W.E., 1926. *Science Service as One Expression of E.W. Scripps's Philosophy of Life*. Washington: Science Service.
Ross, I., 1936. *Ladies of the Press: The Story of Women in Journalism by an Insider*. New York: Harper & Brothers.
Schudson, M., 1978. *Discovering the News: A Social History of American Newspapers*. New York: Basic Books.
Science News Staff, 2022. Some past *Science News* coverage was racist and sexist. We're deeply sorry. *Science News*, 24 Mar., https://www.sciencenews.org/article/past-science-news-coverage-racism-sexism.
Science Service, 1921a. *Science News Bulletin*, 1, 2 Apr., https://centennial.societyforscience.org.
Science Service, 1921b. Big families for well born, urged to rescue race. *Houston Post*, 30 Nov., p. 8.
Science Service, 1924. Immigration selection advocated by eugenists. *Science News-Letter*, 4(145), pp. 7–8.
Science Service, 1932. 800-year wine not fit to drink, professor finds in analysis. *New York Times*, 8 Feb., p. 8.
Science Service, 1933. Eugenicists approve new German law. *The Science News-Letter*, 24(656), p. 295.
Scripps, E.W., 1919. The American society for the dissemination of science. In: Gardner, G. (ed.), History of the Scripps Concern. Ohio University Libraries, E.W. Scripps Papers, MSS 117, Series 4, http://bit.ly/scripps_mss117, 5 Mar., pp. 302–307.
Scripps, E.W., 1920. Notes on the new society. In: Gardner, G. (ed.), History of the Scripps Concern. Ohio University Libraries, E.W. Scripps Papers, MSS 117, Series 4, http://bit.ly/scripps_mss117, 16 Mar., pp. 293–301.
Simpson, C.R., 1987. Nature as news: Science reporting in "The New York Times" 1898 to 1983. *International Journal of Politics, Culture, and Society*, 1(2), pp. 218–241.
Slosson, E.E., n.d. Suggestions as to the work of the science news service. In: Gardner, G. (ed.), History of the Scripps Concern. Ohio University Libraries, E.W. Scripps Papers, MSS 117, Series 4, http://bit.ly/scripps_mss117, pp. 308–319.

Slosson, E.E., 1913. Science and journalism: The opportunity and the need for writers of popular science. *The Independent*, 74(3360), pp. 913–918.

Slosson, E.E., 1919. *Creative Chemistry: Descriptive of Recent Achievements in the Chemical Industries with Many Illustrations*. Garden City: Garden City Publishing Company.

Slosson, E.E., 1921. A new agency for the popularization of science. *Science*, 53(1371), pp. 321–323.

Slosson, E.E., 1922. The writing of popular science. *Science*, 55(1418), pp. 241–242.

Slosson, E.E., 1924a. How women control the future. *Science News-Letter*, 4(144), pp. 6–7.

Slosson, E.E., 1924b. The melting pot. *Science News-Letter*, 5(181), pp. 1–2.

Sponsel, A., 2002. Constructing a 'Revolution in Science': The campaign to promote a favourable reception for the 1919 solar eclipse experiments. *The British Journal for the History of Science*, 35(4), pp. 439–467.

Stafford, J., 1934. Race betterment. In: Davis, W. (ed.), *The Advance of Science*. New York: Doubleday, Doran & Company, pp. 275–283.

Steiner, L., 1997. Gender at work. *Journalism History*, 23(1), pp. 2–12.

Swanberg, S.E., 2019. Psychological armor: The science news-letter warns against propaganda (1926–1965). *Journalism Studies*, 20(13), pp. 1883–1902.

Swanberg, S.E., 2020a. Wounded in mind: Science service writer, Marjorie van de Water, explains World War II military neuropsychiatry to the American public. *Media History*, 26(4), pp. 472–488.

Swanberg, S.E., 2020b. Borrowed chronicles: New York Times science journalist, William L., Atomic Bill Laurence, and the reports of a Hiroshima survivor. In Mays, M. (ed.), *Legacies of the Manhattan Project: Reflections on 75 Years of a Nuclear Age*. Hanford Histories Series Volume II. Pullman: WSU Press.

Swanberg, S.E., 2021. "Well-bred and well-fed," the science news-letter covers eugenics: 1924 to 1966. *American Journalism*, 38(2), pp. 202–230.

Tobey, R.C., 1971. *The American Ideology of National Science, 1919-1930*. Pittsburgh: University of Pittsburgh Press.

Unus, W., 2018. Carr V. Van Anda and the advancement of science news coverage. *Southwestern Mass Communication Journal*, 33(2), http://swecjmc.wp.txstate.edu.

Van Anda, C., 1931. The unsolved riddle of the solar system. *Science*, 74(1912), pp. 187–195.

Wilensky, J., 2021. Cornell's first female doctoral grad: remembering May Preston Slosson, PhD 1880. *Cornellians*, 5 Oct., https://alumni.cornell.edu/cornellians/cornells-first-female-doctoral-grad.

2

THE EXPANDING ROLE OF SCIENCE JOURNALISM

BBC radio in post-war Britain

Jared Robert Keller

Science journalism emerged as a recognised news beat in the first half of the twentieth century when newspapers began to hire specialist writers who could produce engaging stories from increasingly esoteric science (see Chapter 1). In the new medium of radio, however, the role of science journalists took longer to settle. In the UK, as the country emerged from World War II, it was scientists, not journalists, who developed and delivered science programmes for the British Broadcasting Corporation (BBC). Yet by the end of the 1960s, the national radio service had come to prioritise science programmes that were developed and delivered by professional science journalists. This chapter traces the increasing importance and emphasis placed on broadcast science journalism in the period from 1945 to 1970 and examines the BBC's shifting conception of its role as a communicator of science. The debates within the BBC during this period illustrate how multiple factors needed to converge before journalists were entrusted with explaining and presenting science on air.

The development of science broadcasting, and the form it took, reflected both a country's radio culture and its popular science traditions. As one of the first public service broadcasters, the BBC was early to science broadcasting and to the use of specialist science producers, yet relatively late in adopting a journalistic approach to its science output. Before the outbreak of World War II, only eight countries, including the UK, carried science broadcasts. The BBC's position as a publicly funded monopoly with an explicit educational remit meant that it broadcast far more science programming than broadcasters in commercial systems, like those in the United States and Australia, where science broadcasts had to compete with more commercially viable content (LaFollette, 2008). However, for the same reason, the BBC's

DOI: 10.4324/9781003326724-4

pre-war science programming drew on educational, rather than journalistic, formats. This began to change, slowly, in the decades after the war.

In what follows, I use the BBC Written Archives Centre's rich source of archival material on the planning and production of science programming to illustrate the myriad conversations, tensions, deliberations and struggles that brought about the slow acceptance – and then entrenchment – of science journalism within the BBC. I examine these changes by detailing the work of four men involved in the production and delivery of science programming during the post-war period, each of whom embody an important phase in this transition: Archie Clow, a producer of science programming who was wary of journalism and journalistic values; C.L. Boltz, a practising scientist turned professional science writer and journalist; David Wilson, a professional science journalist with little by way of scientific training but with a wealth of journalistic training; and Mick Rhodes, a producer of science programming who championed not just science journalism at the BBC, but a particular style of probing, investigative science journalism. Each phase of this transition shifted power to the growing profession of science journalism, while subtly altering the role that the BBC envisioned for itself when it came to the communication of science.

In part, these changes were the result of an increasing focus on news and journalism across the BBC as a whole from the late 1950s; a time when the growing prominence of television within the BBC's operations was challenging traditional genre boundaries. The corporation began to supplement its reporting with additional journalistic methods such as examination, interpretation and explanation – methods which other broadcasting services in countries like France and the United States had been using for decades. In its science programming, too, the BBC began to embrace more mediated, journalistic formats, which examined, interpreted and explained scientific developments to the British public as a political journalist might explain developments in government policy. Broadcasters increasingly moved away from formats like the straight talk where scientists spoke directly to listeners, towards formats like interviews, documentaries and news-style reports which enabled BBC journalists and programme developers to contextualise – and even criticise – what scientists said. As such, the role that the BBC envisioned for itself shifted from the *presentation* of scientists to the British public to the *mediation* of science and scientists on behalf of the British public. As the BBC expanded its role, scientists – who had previously enjoyed a seemingly direct connection to listeners – found themselves increasingly relegated to a position one step removed from listeners; cast as contributors to science programming rather than as deliverers of science programming.

BBC journalism before 1945

For the first decades after its creation in 1922, the BBC had maintained a strict demarcation between 'news', which strove to present the bare facts, and 'comment', which had greater freedom to analyse and interpret. By separating news from comment, the BBC's first Director General, John Reith, had sought to reassure the British press that the new broadcasting organisation posed no threat to the future of the print media, while simultaneously assuring the British government that it could be trusted to remain impartial and would refrain from offering its own editorial opinion on controversial subjects.

This demarcation between news and comment became ingrained in the very structure of the BBC with the creation in 1935 of two distinct departments: the News department and the Talks department (Chignell, 2009: p. 152). The News department aimed to put out unbiased, objective news and information; the Talks department focused on comment, opinion and analysis programmes. This demarcation proved a constant point of struggle over the first 30 years of the BBC and greatly influenced the content, format and framing of science broadcasting over that period. Other BBC departments such as Schools and Features also produced science programming during this time (Jones, 2013: p. 441), but it is the evolving relationship between the Talks and News departments that shaped the development of radio science journalism in the UK.

Although the BBC's emphasis on impartiality and neutrality mirrored the 'facts-only' style of newspaper journalism that had emerged in 1920s America and, less formally, in Britain (see chapters 1 and 8), there was widespread suspicion of journalists and news values in the early BBC. As literary critic John Carey has noted, the cultural elite of Britain – from which the early BBC's senior positions were often filled – had long exhibited a disdain for journalists and their ilk (1992: p. 7). True to type, in 1927 the head of the BBC's first news section, Geoffrey Strutt, had expressed a desire to steer clear of the "errors into which journalists, as such, seem inevitably to fall (sensationalism, inaccuracy, partiality and overstatement)" (Scannell and Cardiff, 1991: p. 113).

This distrust of journalists, combined with the separation of news from comment, resulted in News department programmes with a reputation for being rather staid – especially when compared to the news broadcasts of other countries where there was no clear separation between news and comment. For instance, in the years following World War II, the contemporary US broadcasting scholar, Burton Paulu, judged that there was too little news analysis on the BBC: "One turns to BBC news programmes for news and nothing else!" (Briggs, 1995a: p. 522). American broadcasters like Edward R. Murrow, who had impressed Americans and Britons alike with his war-time reports from London, were often involved in the researching and writing of their own material and had the freedom to express their own personal views within a broadcast. By contrast, Tahu Hole, editor of BBC News in the early 1950s, maintained that it would be disastrous to "Americanize" the news by letting newsreaders develop their own material or offer their own personal views (Briggs, 1995b: p. 69). As historian Hugh Chignell points out, at that time 'Americanize' was "one of the worst slurs that could be used against the BBC" (2009: p. 32).

On the other side of the news-comment divide, the BBC's desire to remain impartial meant that when engaging with potentially controversial topics, the early Talks department almost exclusively turned to outside experts. By recruiting prominent thinkers from various political, ideological and philosophical backgrounds to voice their opinions on potentially controversial matters, the BBC could ensure that any controversial opinions or comments would emanate from outside experts rather than from the BBC itself.

The forced separation of news from comment, the widespread distrust of both journalists and journalistic values and the preference for recruiting prominent external voices all greatly influenced the shape of the corporation's science coverage during the early post-war period. The work of four individuals spanning the quarter of a century after the war illustrates how the BBC's attitude towards science journalists gradually changed in an evolving media landscape where radio increasingly needed to compete with the new medium of television.

"Scientists themselves"

Archie Clow, the first of the four men who help to illustrate the BBC's transition towards professional science journalism, joined the BBC in July 1945 as a producer and programme-developer. He went on to become Editor, Science Talks (Radio), producing over 2,500 science programmes during his 25-year career at the BBC (Keller, 2017). Clow was an adept communicator of science, having been a founding member of the Aberdeen Scientific Film Club (one of the first of its kind in Britain) and co-author, with his wife Nan Clow, of a 680-page book on the history of industrial chemistry (Clow and Clow, 1952). However, his role when he joined the BBC was not to communicate science; rather, it was to help *scientists* communicate science, even if this meant ceding control over the framing and content of radio science broadcasts.

On the night of 21 March 1956, 11 years into his career at the BBC, Clow attended a conference at the Senate House Library entitled 'The Presentation of Science to the Public'. According to a report on the conference, it was becoming "increasingly important that members of the public should understand in general terms what the scientists are doing and how the results of their work are likely to affect daily life." As the producer of *Science Survey* – a long-running science series composed of talks by eminent scientists – Clow had been invited to speak on "the presentation of science by sound radio" (RIC, 1956).

Clow began the proceedings that night by describing himself as a "scientific impresario" (Clow, 1956: p. 438). When developing a science talk, he explained, he had "an obligation to present the scientist himself: to bring the authority to the microphone and present him to the public" (p. 437). He understood his role as a radio science producer to involve the transformation of the scientist into a radio presenter: "the presentation of science *through the medium of the scientist himself* is my problem – and being presented that of the scientist"

(p. 437, emphasis added). Solving this problem was a notable achievement; thus when *Science Survey* had first launched, the promotional material for the series boasted that: "For the first time in British radio history there is to be a long and regular weekly period in which scientists themselves will talk about their work and current topics" (BBC, 1946).

Clow's job was to stay abreast of developments in science that were likely to be of interest to listeners, identify authoritative scientists working in those fields who were capable of producing a satisfactory talk, entice those scientists to make an appearance on BBC radio, prepare them for the broadcast, advertise the performance and attract an audience. He could advise scientists on the presentation of a talk (the structure, the language or the overall style, for instance) and he might even exert an element of control over the framing of the talk (such as which ideas to stress, which lines of argument to take, and so on), but the actual content of the talk would be left almost entirely to the scientist. Indeed, although Clow told the conference attendees that "the best broadcasts are the products of close co-operation between scientist and producer", he clarified that in such an effort, "the facts are the responsibility of the scientist: the treatment is open to discussion" (1956: p. 438).

Opting to use scientists to deliver information about their own fields did not in itself restrict the BBC's science programming to lecture-style talks – science producers could have opted for discussions or interviews, for instance – but, as Clow explained at the same conference, his department's emphasis on the facts of science led "naturally" to the talks format, since "the question of opinion, the mainstay of discussion, does not enter very often" (1956: p. 437). Throughout this period, then, the aim was not to develop programmes that would interpret science, provide important context or explain the hidden biases and complications woven throughout a science-related issue; rather, it was to provide a platform from which scientists could inform listeners about important developments as they saw fit. Clow could, naturally, still frame the content of a talk through his choice of speaker and could exert a measure of control over the final talk by offering suggestions on a scientist's initial drafts, but ultimately his role concerned presentation rather than interpretation or explanation.

This approach was in keeping with Talks department policy as a whole at the time. As George Barnes, the BBC's Director of the Spoken Word, explained at a 1949 meeting of the General Advisory Council:

> The purpose of Talks Department's work is to keep the public informed of developments in those departments of knowledge which can be made interesting to listeners through sound broadcasting and this is done by the producer finding the most appropriate and authoritative speaker on the subject selected.
>
> (Barnes, 1949)

The preference for authoritative external speakers was in part due to the BBC's desire to remain impartial, but it was also a way of claiming status for the BBC and securing a position for itself among Britain's elite cultural institutions (Jones, 2010). It also appears to have been a result of the fact that science producers like Archie Clow shared the BBC's deeply ingrained suspicion of 'news men' and journalism – particularly American-style journalism. Indeed, in 1947, W.M. Newton, the assistant director of Talks, had written to Clow to inform him that although he had been receiving the scripts of science programmes broadcast by the American network CBS, he was disappointed to find that they were "diluted with journalism to a degree" (Newton, 1947). Clow seems to have agreed that American output compared unfavourably with his own programmes, noting in a letter of the following year that in the United States "there doesn't seem to be a steady output of, I hope I can use the adjective, reputable, broadcasts like *Science Survey*" (Clow, 1948).

It was not simply that practising scientists were deemed more prestigious and knowledgeable than science journalists. Scientists, it was assumed, would communicate their research in the sober, responsible manner desired by the BBC. Journalists, on the other hand, were wont to exaggerate, cause misunderstandings, take narrative and linguistic liberties, and possibly even lead the public to adopt negative or unrealistic expectations of science. For the first decade after the war, official BBC policy was to avoid such a dangerous approach to science on its domestic services. The preference was for scientists to communicate news about science themselves and for the BBC to serve as a platform from which those scientists could speak. Science journalists were, as a rule, not welcome. However, this was about to change.

The arrival of science journalists

Up until the late 1950s, the BBC's domestic radio services had never employed a science journalist. In fact, the domestic services as a whole rarely put journalists or correspondents of any sort on air. This began to change with the arrival in 1955 of the first commercial television network, ITV, which meant that BBC News now needed to compete for the domestic audience. In his autobiography, Ian Trethowan, who later became Director-General of the BBC, described the situation:

> The idea of an experienced specialist journalist explaining the events, or interviewing one of the participants to try to elicit more from him, was only grudgingly accepted, and mainly from abroad, through the BBC's excellent foreign correspondents.
>
> (Trethowan, 1984: p. 80)

The overseas services, having always been exposed to market competition, did not share the domestic service's distrust of journalists and journalism or the strict demarcation between news and comment. Indeed, the BBC's European Service had been making use of science journalists since the late 1940s, to much acclaim. By the late 1950s, influenced by the popularity and success of this more journalistic overseas approach, there was a growing appetite within the BBC domestic services for journalists to be heard on air and for the broadcasting of views and comments as well as facts and information (Briggs, 1995b: p. 547).

The corporation also considered the question of what news it should cover. A 1957 survey of audience attitudes towards the BBC's news services indicated high levels of interest in news about science: the most requested news items were pieces on "accidents and disasters", followed by news about "science and technology", with news reports on the weather coming in third (Briggs, 1995b: p. 71). Within two years, the BBC had appointed a 'science correspondent' to meet this demand for science news: C.L. Boltz.

Boltz is the second of the four men who help to illustrate the BBC's transition towards professional science journalism. He had begun his career as a research scientist and lecturer before becoming a science journalist at the BBC's European Service and General Overseas Service for nearly a decade. His move from the overseas to the domestic services in the spring of 1959 can be seen as the beginning of a shift in control of BBC science coverage from scientist to journalist.

Boltz's appointment to the domestic services met with approval from the scientific community. Indeed, the journal *Nature* suggested that the domestic services could not have found a better person. "Few people have had so wide an experience of science broadcasting", a report on Boltz's new role noted:

> for during the time he has been broadcasting from Bush House he has covered a vast range of scientific topics. Among European listeners he has a great reputation and he is one of the most popular broadcasters to give talks at Bush House.
>
> (Nature, 1959)

Within these overseas broadcasts, Boltz appears to have taken – and, perhaps more importantly, was allowed to take – comparatively more journalistic, narrative and linguistic liberties than did the scientists who appeared on the domestic services. For instance, Boltz kicked off a 1956 news report with the rather startling pronouncement: "I have to admit that I am a drug addict. What is more – so are you." Tongue planted firmly in cheek, he continued:

> At this very moment thousands – perhaps millions – of British people are taking the dried and bruised and broken leaves of a sort of oriental camellia

> and making from them an infusion with boiling water to dissolve out the drug. In other words, they are making tea.
>
> (Boltz, 1956: p. 1)

Such playful passages were commonplace in the scripts of many of the science correspondents on the overseas services.

During the hiring process for the new science correspondent, Archie Clow foresaw that the new position represented a shift in the BBC's approach to domestic science programming. At a 1958 meeting of BBC personnel involved in the production of science programming, Clow cautioned that the appointment of a science correspondent would "represent a departure from the precedent that science broadcasts should be provided by scientists themselves" (Shaw, 1958: p. 1). Speaking later at the same meeting, Boltz – who had not yet been selected for the position – commented that in his mind such a departure could be achieved with very little difficulty. According to the minutes of the meeting, Boltz advised his colleagues that "almost any producer could deal with a scientific subject without any obligation to refer for advice to experts who might recommend the right speakers or the right treatment" (Shaw, 1958: p. 2). Here, then, was someone who did not feel constrained by three decades of BBC precedent. For Boltz – and others like him – the production, planning, writing and even *delivery* of science programmes could be understood as falling in the domain of the professional science journalist, producer or broadcaster.

As the BBC moved into the 1960s, the idea that science programmes could be delivered by science journalists and commentators rather than exclusively by scientists themselves was slowly catching on. Importantly, in supporting this new approach, the BBC was extending its existing control over access to the airwaves and the framing of a broadcast, to control over the on-air delivery of broadcasts. By delivering content through its own personnel, rather than continuing to delegate this task to scientists, the BBC was positioning itself as a communicator of science in its own right: no longer just a platform for scientists, the BBC was starting to mediate science and scientists on behalf of the British public.

A particular type of science journalist

If the late 1950s witnessed the slow acceptance of both journalism and science journalists within the BBC, the early 1960s witnessed the development of a preference for a particular *type* of science journalist – one that was a journalist first and need not have a scientific training. Indeed, despite the fanfare accompanying Boltz's appointment as science correspondent in 1959, the BBC of this period was still working out how best to utilise science journalists and what it expected from them. By the autumn of 1960, Boltz had resigned his post, telling his colleagues that in his opinion: "the news outlets for science material did not justify a full time Science Correspondent" (BBC, 1960). The position sat empty for over two years, until in the autumn of 1962, J.A. Camacho, Head of Talks and Current Affairs (Sound), informed his superiors that he intended to "re-establish the post of Science Correspondent" (1962: p. 3). By the following spring, Camacho had hired David Wilson to fill the post – despite the fact that, unlike Boltz, Wilson was not a trained scientist. Wilson, the third illustration of the BBC's transition towards professional science journalism, represents a growing prioritisation of journalistic rather than scientific training.

Wilson had, in fact, briefly worked towards a mathematics degree at Cambridge before departing for national service with the Royal Artillery in Sri Lanka (Adeney, 2000). Upon his return to Cambridge, he switched his focus to history and began honing his journalistic skills at the student newspaper. As Camacho put it, it was only after working at the BBC as a "general reporter" for some time that Wilson began "specialising in science stories" around the summer of 1962 (Camacho, 1962: p. 3).

It was Wilson's proven record as a news journalist that had most recommended him to Camacho. Indeed, Camacho had disagreed with Boltz's assessment in 1960 that there simply were not enough news outlets to justify a full-time Science Correspondent. For Camacho, Boltz's scientific training had been a liability rather than an asset, preventing him from making the most of the opportunities for science reporting. "The experiment tried some years ago of appointing a scientist as Science Correspondent in the News Division was not successful substantially", Camacho wrote in his memo making the case for reintroducing the post, because Boltz, "though well respected in the scientific world, found it difficult to adjust himself to the needs of news" (1962: p. 3). By contrast, Camacho saw "no risk" in appointing Wilson, since he had already shown himself to be "capable of adjusting his work to the needs of News Division." In Camacho's opinion, experience as a journalist trumped scientific training.

The hiring of Wilson in 1962 signalled a concerted effort on the part of the BBC to increase the coverage of science in its news programming. Indeed, in May 1964, at the first meeting of the Science Consultative Group – a twice annual meeting between elite scientists and BBC staff – the BBC argued that the inclusion of science in news and topical programmes was "perhaps the surest way" of reaching the largest audience possible. Furthermore, it was noted, "some expansion in this direction is planned" (BBC, 1964: p. 2).

Later that year, the BBC hired a 'Science Reporter' to assist Wilson, and by November 1965, the BBC could report back to the Science Consultative Group that news reports on science had indeed "expanded markedly in recent years" (BBC, 1965: p. 2). Importantly, these news reports on science were more often than not delivered by science journalists who prioritised journalistic skills and expertise over scientific knowledge and training.

By the mid-1960s, the BBC had begun to prefer this type of science journalist. In a 'lunch-time lecture' at Broadcasting House in February 1966, Aubrey Singer, then Head of Outside Broadcasts, Feature and Science Programmes (Television), confirmed this prioritisation of journalistic skills and values when he asserted that whether or not a broadcaster had undertaken scientific training was "fundamentally unimportant" (1966: p. 8). Singer described the work of his department as "science journalism", referring to the producers in his department as "creative and conscientious journalists" and praising their "individual journalistic initiative" (pp. 8, 13, 12). His lecture that day was nominally about television, but Singer – who would go on to serve as Managing Director of Radio – was also careful to point out that in so far as the lecture touched on "broad aspects of mass communication," it could be "related to broadcasting generally" (p. 3).

A copy of Singer's lecture was sent to the members of the Science Consultative Group. The minutes of their following meeting record that David Wilson, "in support of Mr. Singer's thesis", pointed out to the scientists present that "many newspaper science correspondents, including some of the best ones, were journalists without any scientific training." (BBC, 1966: p. 6). In stark contrast to the suspicion towards journalism just a decade previously, by the mid-1960s, at least some at the BBC had come to prioritise journalistic values over scientific values and the needs of news over the needs of science.

A particular type of science journalism

In the years that followed, the BBC would come to prefer not only a particular type of science journalist, but a particular type of science journalism: one that was probing, critical and investigative. In his 1966 lunch-time lecture, Aubrey Singer put this ethos into words most clearly and forcefully. "The aim of scientific broadcasting", he explained:

> is to create a climate of critical opinion, in which the work of the scientist can be understood, encouraged, and criticized, so that the body politic can intelligently share in and influence the course of science […].
>
> (Singer, 1966: p. 15)

Science broadcasting, Singer believed, had "a vital role" to play in fostering this climate. The inclusion of a critical stance was significant and was to define the BBC's approach to science programming for the rest of the decade.

Similar changes were occurring in science journalism generally. Historians have identified a shift during this period away from positive and deferential coverage of science towards a more critical form of science journalism (Bauer and Gregory, 2008; Nelkin, 1987). Whereas science journalists of previous decades had tended to think of themselves as advocates for science and conveyers of scientific information, by the late 1960s they were beginning to see themselves as professional journalists with a duty to cast upon science the same critical gaze which their colleagues working on other beats routinely cast upon politics or culture. For instance, Henry Pierce, a science reporter at the *Pittsburgh Post Gazette*, observed in 1966 that while other journalists maintained a healthy scepticism towards their news sources, "we, bless us, go in with our bright baby-blue pencils poised, faithfully recording anything our scientists – gods – tell us. Never does it occur to us that these guys too may have motives that are less than noble" (Nelkin, 1987: p. 89). Increasingly, the deference that science journalists had traditionally extended towards scientists disappeared and a commitment to journalistic independence took its place. With this new ethical commitment to independence and scepticism came a willingness – perhaps even a responsibility – to reflect contemporary public concerns about science and technology. Science, in the words of many at the time, had been politicised.

This commitment to reflecting public concerns also fitted with the BBC's approach to journalism as a whole during this period. By the mid-1960s, producers and programme-developers stressed the need for the BBC to engage with the social climate of the time, characterised by civil rights protests, counter-culture movements and environmentalism (Mills, 2016). In 1965, for example, Hugh Greene, then Director-General of the BBC, declared that in his opinion the role of the BBC was to "encourage the examination of views and opinion in an atmosphere of healthy scepticism" (Hollinshead, 2002: p. 75). This emphasis on scepticism led to the development of 'current affairs' programmes that melded news and comment and had a much more journalistic feel – a development which would have been unthinkable in previous decades during the forced separation of news and comment.

The series that perhaps most embraced and reflected this changing approach to news and journalism during the latter half of the 1960s was the radio news series, *The World at One*, which was inspired by the informal, conversational style of the newsreaders on French public radio (Chignell, 2011; Hendy, 2007). William Hardcastle, the man selected to anchor *The World at One*, was a Fleet Street journalist who fully embraced new methods of presentation such as radio actuality inserts, interviews, reports from correspondents on the ground and phone interviews with witnesses to important events. Hardcastle even developed a probing, and at times confrontational and combative, style of interview. The series was so popular that its combination of informality and hard-hitting journalism began to be copied by producers and programme-developers throughout the BBC. As Sue MacGregor, the presenter of the radio magazine *Woman's Hour*, put it: "I think many of us began to copy Bill – however unconsciously" (Hendy, 2007: p. 49). According to media historian David Hendy, the widespread imitation and duplication of *The World at One* represented "more than an act of colonization for journalism *per se*. It also represented the growing preference within the BBC for a particular *kind* of journalism" (2007: p. 50): a harder-edged, more probing, less reverential style of journalism.

As programmes like *The World at One* merged news and comment, they also altered the very purpose of news programmes and the role the BBC envisioned for itself within British society. Whereas the BBC's early news services had traditionally defined their role as *informing* the audience about what was happening, the new current affairs programmes defined their role as helping the audience *understand* what was happening (Chignell, 2009: p. 151).

These two trends – the rise in many countries of a more critical form of science journalism and the desire within the BBC for probing journalism rooted in 'healthy scepticism' – led to yet more changes in how science was presented on the BBC and in how both science producers and science journalists at the BBC conceived of their roles. Michael 'Mick' Rhodes, the final illustration of the BBC's transition towards professional science journalism, demonstrates the impact of both of these trends on the science programmes produced by the Talks department during this period. Having joined the Talks department as a science producer and programme-developer in the mid-1960s, Rhodes was part of the new guard that began to refashion BBC radio programming during the late 1960s. As such, he stands in contrast to his boss, Archie Clow, whose two decades at the BBC had left him steeped in the corporation's traditional culture.

By the time Rhodes celebrated his third anniversary in the Talks department in 1967, he had developed a probing, investigative, interpretive, and at times critical approach to science broadcasting. Importantly, he frequently included reports by science journalists within his programmes and actively sought to craft programmes that engaged with and reflected the temper of the times. A case in point is *New Worlds*, a series that Rhodes intended to be a direct response to the changes within both society and the BBC. It was to be more youthful, more conversational and more 'newsy' while being less ponderous, less deferential and less scripted than science news series of previous decades. Most importantly, Rhodes intended to utilise correspondents and journalists to a greater degree than ever before. During the planning of the series, Rhodes was explicit about this: "Clearly the programme would use scientific journalists much more than we have in the past" (1966: p. 2). In this, Rhodes was directly challenging Talks department policy as it had existed since the end of the war and was signalling his willingness – indeed determination – to move beyond the edict to exclusively use scientists in the delivery of science programmes. What is more, he wanted the science journalists who appeared in his series to have the freedom to critically investigate science, rather than serve as passive communicators of what scientists wanted to say. The facts of science would no longer be the sole domain of scientists.

From his very first memo, Rhodes was clear that he sought to create a weekly "in-one-place science news service" unlike anything that had existed on BBC radio before (1966: p. 2). This prioritising of up-to-the-minute news mirrored the developments in news and current affairs brought about by programmes like *The World at One*. Rhodes aimed to craft a similarly modern news series that could report on important or pressing topics involving science, through a combination of interviews, discussions, actuality recordings, reports from scientists, investigative reports from science journalists, and commentaries and explanations from BBC personnel (BBC, 1967: p. 5). Through investigative reports and interviews, Rhodes hoped his *New Worlds* series would emulate the "cutting William Hardcastle postscripts" for which *The World at One* had become famous. Indeed, the creator of *The World at One*, Gerard Mansell, was fully supportive of Rhodes' plans for the innovative science news series. "I've read your proposal for a new weekly Science Magazine [...] with very great interest and am in particularly close agreement with what you say about style and target audience", he wrote to Rhodes. "I am very much in favour of what you propose and hope we can get a really lively, popular programme of science news and comment off the ground fairly rapidly" (Mansell, 1966).

Thus, when *New Worlds* debuted on 20 October 1967, it did not represent the arrival of something wholly new on BBC radio. Rather, it represented the continuation and outgrowth of much that had been brewing over the post-war period, both within science programming – through people like C.L. Boltz and David Wilson – and across the BBC more generally – such as the team behind *The World at One*. And yet, although the debut of *New Worlds* did not represent something new on BBC radio, it did represent something of a takeover for science journalism. The same year that *New Worlds* debuted, three long-running science series that had been developed or stewarded by Archie Clow were cancelled: *Science Survey*, *Science Review* and *Who Knows?*. As a result, *New Worlds*, was poised to become *the* major science series on BBC radio. Indeed, from its debut in 1967, *New Worlds* was the most visible and arguably most important science series on BBC radio until it ended in 1973.

As the 1960s came to a close, the BBC's major science series was organised around reports by professional science journalists and was motivated by a desire to bring science out into the open so the public could debate its merits and potential drawbacks. Like *The World at One* and other news and current affairs programmes during the late 1960s, *New Worlds* set itself up as a programme that could not only inform people, but also help them understand. Whereas Clow had seen his role within *Science Survey* as the communication of science 'through the medium of scientists themselves', Mick Rhodes saw his role with *New Worlds* as the examination and critical investigation of science through journalistic reports by professional science journalists.

Science journalists had by no means completely replaced scientists on the BBC – scientists still appeared in interviews and discussion programmes and delivered the occasional straight talk – but science journalists had become an integral part of the BBC's approach to science broadcasting. Just how important science journalists were perceived to be for the future of BBC radio can be seen by contrasting the launch of *Science Survey* in 1946 with the launch of *New Worlds* in 1967. To celebrate the launch of *Science Survey*, the BBC had hosted a celebratory cocktail party for eminent scientists, the guest list to which included two dozen of Britain's top scientists (Cox, 1946). Two decades later, to celebrate the launch of *New Worlds*, the BBC once again hosted a celebratory cocktail party – but this time for journalists. In an August 1967 letter to management, Rhodes requested "some sort of publicity occasion for science journalists at which we could say what we are trying to do. This presumably would mean a playback and a certain amount of alcoholic reinforcement" (Rhodes, 1967). The BBC's files for *New Worlds* contain no mention of whether a comparable amount of alcoholic reinforcement was also provided for scientists, but the importance Rhodes placed on the event for journalists shows the extent to which the BBC felt it needed to cultivate science journalists as future contributors.

Conclusion

As science journalism became established in the first half of the twentieth century, it was not inevitable that the new medium of radio would use science journalists to communicate science. Indeed, the early BBC preferred to use practising scientists in its broadcasts – the result of its explicit educational remit, a desire to avoid controversial topics that might jeopardise its fragile relationship with both the government and the press, and a widespread distrust of journalists and journalistic values throughout the upper levels of the corporation. Archie Clow exemplifies this distrust of journalism within the BBC during the first decade after World War II and the commitment to working through the medium of scientists themselves.

Eventually, the distrust of journalism began to fade – in science programming circles as well as on the news side of the BBC. Staff within the corporation were influenced by the different approach to journalism in countries like the United States and France, where news presenters were intimately involved in programme production. They were also influenced by the popularity and success of the BBC's own overseas broadcasts, which had much earlier adopted a more journalistic approach. As the distrust faded, science journalists who had been plying their trade on the overseas services at the BBC began to carve out a place for themselves on the domestic airwaves as well. The hiring of C.L. Boltz represented a shift towards the BBC becoming a mediator and communicator of science rather than a presenter of scientists. The following years saw the increasing professionalisation of science journalism and an increasing focus on journalistic, rather than scientific, training and expertise, as exemplified by the science journalist David Wilson.

Even so, it also wasn't inevitable that radio science journalists should seek to interrogate and problematise science. Even in countries like the United States where broadcasters had earlier used popularisers and journalists in preference to practising scientists, these popularisers were still primarily advocates for scientists rather than critical interrogators. It took wider changes in the professional ethos of science journalism, as well as changing cultural attitudes towards authority during the counter-culture movements of the 1960s, for science journalists to begin to see their role as the interrogation and problematisation of science. Science journalists shared the changing values of journalist colleagues on other beats and the move to what journalism scholars Katherine Fink and Michael Schudson (2014) have called "contextual reporting". At the BBC, science producers and programme developers like Mick Rhodes were inspired by the hard-hitting journalism of their peers elsewhere in the corporation and sought to emulate it within their own programmes. As they did so, they reinterpreted their role: no longer just science journalists but probing, critical science journalists.

The transition was by no means quick, but the BBC slowly moved from prioritising straight talks by scientists themselves, to talks by science journalists with scientific training, to news-style reports by science journalists with journalistic training, and finally to news-style reports by science journalists with a probing, investigative style. By the late 1960s, the BBC had reconceived its role as a communicator of science in its own right. As the BBC found its own voice, its producers, programme-developers and science journalists began to see their job as the examination, interpretation and explanation of scientific developments to the British public. They increasingly presented science through interviews, news-style reports and documentaries – formats which placed broadcasters and science journalists in a position to mediate science by explaining, contextualising and summarising what scientists said.

As the BBC embraced this more mediatory role, scientists were increasingly consigned to the role of contributor – thereby losing the direct connection to listeners they had once enjoyed and much of the control they had exercised over the public image of science. In 1956, Clow had explained to his audience at 'The Presentation of Science to the Public' conference that "the facts are the responsibility of the scientist: the treatment is open to discussion." (p. 438). By the late 1960s, the facts were no longer the sole responsibility of the scientist. Producers and science journalists were in a position to choose which facts were relevant to a given issue as they contextualised and questioned scientists' contributions.

It took decades of science broadcasting for science journalism's role as a mediator to emerge and for science journalists' professional expertise to be seen as including the ability to explain and provide context for scientific developments. Even then, the role of science journalism was not settled and continued to evolve. Indeed, the rise in professional public relations for science over the last few decades can be seen as the latest iteration of scientists' attempts to regain the control they once enjoyed in their radio broadcasts. In the post-war period, broadcasters worked in the context of changing media technologies and changing cultural values, including a less deferential public sphere and an increasing awareness of the social impact of science. The historical example of the BBC's radio programming in the post-war period reminds us that the place of journalism within a given medium is the product of complex contextual factors such as these. In the 1960s, professional science journalists found a new role at the BBC as the organisation adapted in light of wider social developments. Over half a century later, in a new media landscape, science journalists' roles within media organisations and their relationships with the institutions of science continue to be redrawn in complex and culturally specific ways.

References

Adeney, M., 2000. David Wilson. *The Guardian*, 19 Sep., http://www.theguardian.com/news/2000/sep/19/guardianobituaries2.

Barnes, G.R., 1949. Broadcasts on Science – Notes by Mr. G.R. Barnes, 4 Jul. Written Archives Centre, R6/34

Bauer, M.W. and Gregory, J., 2008. From journalism to corporate communication in post-war Britain. In: Bauer, M.W. and Bucchi, M. (eds.), *Journalism, Science and Society: Science Communication between News and Public Relations*. London: Routledge, pp. 33–51.

BBC, 1946. Science Survey. Memorandum. BBC Written Archives Centre, R51/533/1.

BBC, 1960. Extract from the minutes of D.G.'s news and current affairs management meeting held on Friday, 23rd Sep. 1960. BBC Written Archives Centre, R6/239/1.

BBC, 1964. Science Consultative Group – broadcasting about science and technology. BBC Written Archives Centre, R6/239/2.

BBC, 1965. Science Consultative Group – radio coverage of the British Association Annual Meeting. 10 Nov., BBC Written Archives Centre, R6/289/1.

BBC, 1966. Science Consultative Group – note of a meeting held on Thursday, 2 Jun. 1966. BBC Written Archives Centre, R78/76/1.

BBC, 1967. Science Consultative Group – note of a meeting held on Thursday, 8 Jun. 1967. BBC Written Archives Centre, R78/76/1.

Boltz, C.L., 1956. Tea – the national drug. Microfilm Script. *European Service – General News Talk: General Overseas Service*, 21 Aug. Radio Talks Scripts Pre-1970 (Film T462). BBC Written Archives Centre.

Briggs, A., 1995a. *Sound and Vision. Vol. 4. The History of Broadcasting in the United Kingdom*. Oxford: Oxford University Press.

Briggs, A., 1995b. *Competition. Vol. 5. The History of Broadcasting in the United Kingdom*. Oxford: Oxford University Press.

Camacho, J.A., 1962. Science broadcasts. 14 Nov. BBC Written Archives Centre, R6/239/1.

Carey, J., 1992. *The Intellectuals and the Masses: Pride and Prejudice among the Literary Intelligentsia, 1880-1939*. London; Boston: Faber and Faber.

Chignell, H., 2009. *Key Concepts in Radio Studies*. London: Sage.

Chignell, H., 2011. *Public Issue Radio: Talks, News and Current Affairs in the Twentieth Century*. Basingstoke: Palgrave Macmillan.

Clow, A., 1948. Archie Clow to W.T. Astbury. Letter, 6 Jan. BBC Written Archives Centre, RCONT1 Astbury.

Clow, A., 1956. Sound radio. *Journal of the Royal Institute of Chemistry*, 80, pp. 437–442.

Clow, A. and Clow, N.L., 1952. *The Chemical Revolution: A Contribution to Social Technology*. London: Batchworth Press.

Cox, I., 1946. Ian Cox to Head of Light Programme, Weekly half-hour science period – Light Programme. 15 Mar. BBC Written Archives Centre, R51/533/1.

Fink, K. and Schudson, M., 2014. The rise of contextual journalism, 1950s-2000s. *Journalism*, 15(1), pp. 3–20.

Hendy, D., 2007. *Life on Air: A History of Radio Four*. Oxford: Oxford University Press.

Hollinshead, A.M., 2002. *The Production and Reception of Discourses Concerning Religion in Fictional Broadcasting*. PhD Thesis, University of Glasgow.

Jones, A., 2010. *Speaking of Science*. PhD thesis, University College London.

Jones, A., 2013. Clogging the machinery: The BBC's experiment in science coordination, 1949-1953. *Media History*, 19(4), pp. 436–449.

Keller, J.R., 2017. *A Scientific Impresario: Archie Clow, Science Communication and BBC Radio, 1945–1970*. PhD thesis, Imperial College London.

LaFollette, M.C., 2008. *Science on the Air: Popularizers and Personalities on Radio and Early Television*. Chicago: University of Chicago Press.

Mansell, G.E.H., 1966. G.E.H. Mansell to Mick Rhodes, Weekly science magazine. Memorandum, 17 Nov. BBC Written Archives Centre, R51/1269/1.

Mills, T., 2016. *The BBC: Myth of a Public Service*. London: Verso.

Nature, 1959. Science correspondent of the B.B.C.: Mr. C.L. Boltz. *Nature*, 183, p. 1231.

Nelkin, D., 1987. *Selling Science: How the Press Covers Science and Technology*. New York: W.H. Freeman.

Newton, W.M., 1947. Handwritten note on 'Putting science on air'. BBC Written Archives Centre, R51/523/6.

Rhodes, M., 1966. Mick Rhodes to H.T.C.A.(S), Weekly science magazine. Memorandum, 15 Nov. BBC Written Archives Centre, R51/1269/1.

Rhodes, M., 1967. Mick Rhodes to C.H.S. and M.P., Magazine Programme – Monday Through Friday. Memorandum, 23 Aug. BBC Written Archives Centre, R51/1269/1.

RIC, 1956. The presentation of science to the public. *Journal of the Royal Institute of Chemistry*, 80, p. 437.
Scannell, P. and Cardiff, D., 1991. *A Social History of British Broadcasting. Volume 1: 1922-1939. Serving the Nation*. Oxford: Basil Blackwell.
Shaw, C.D., 1958. Science broadcasts – notes of a meeting on Wednesday, 22nd Oct., 1958. BBC Written Archives Centre, R51/967/1.
Singer, A., 1966. Science Broadcasting. *BBC Lunch-Time Lectures: Series 4*. London: British Broadcasting Corporation.
Trethowan, I., 1984. *Split Screen*. London: Hamish Hamilton.

3

CONSTRUCTING IDENTITY, PROTECTING INDEPENDENCE

Science journalists' associations

Jane Gregory

Humans have a tendency to form associations. They have joined together in more-or-less formal, more-or-less rule-bound groups at all stages of human history and all around the world, in both rural and industrial environments (Anderson, 1971). Associations tend to emerge at times of stress or to relieve tension arising from an issue of scale. They sit at an intermediate position, between the individual and the state, for example, or between the family and society. In doing so, they take the individual – or 'private' person – into a more public realm (Gregory, 2004).

Science journalists have shared this human tendency to associate. Since the formation of the National Association of Science Writers in the United States in 1934, other nations and territories have taken the same step. Resolving an issue of global scale, the World Federation of Science Journalists (WFSJ) was formed in 2002 and today claims membership from 67 nations (WFSJ, 2023). This chapter explores the role of such associations in providing science journalists with both professional resources and a collective identity.

The emergence of associations is an indicator of social change. In the West, the twentieth century was the century of the association, with economic crises and the ends of wars (but not their beginnings) being the catalysts for frantic associating. As anthropologist Robert Anderson argued, the procedures and rules of associations give them the capacity "to facilitate the transition of individuals and societies to participation in the modern world" (1971: p. 216). When the world is changing, associations help individuals to adapt to new conditions.

In the English-speaking world and in the North, the 1970s appear to have hosted a final burst of physical association-forming before neoliberalism and the internet changed everything. The decline in associating in these regions has been described as a crisis of civil society and a hollowing out of public life,

DOI: 10.4324/9781003326724-5

suggesting that associations play an important role not just for their members, but also in providing a 'middle way' for wider society (see, for example, Osborne, 2021). However, while the urge to join country clubs, political parties and trade unions may have been lost, science journalists' associations continue to thrive. The theory of associations provides some explanation: one of the key rationales of associations is the perennial belief that 'united we stand, divided we fall'. For individuals who feel oppressed or threatened, there is strength to be found in association, and with it the possibility of economic and political power.

Like a Habermasian public (Calhoun, 1992), associations of private individuals may serve as a counterbalance to the prevailing order. Yet they do so by mimicking the mechanisms of the society in which they are formed: in a democratic society, associations will be democratic. In this way, associations act as a training ground for effective action in the wider social sphere. In Anderson's words, an association functions to "adapt individuals for modern participation" (1971: p. 218). For a science journalist who moves from the newsroom to the lab to the parliament to the launch pad to the earthquake zone, the association provides a point of connection between these otherwise distant sites. This is especially helpful for journalists who are impaired by the notoriously narrow and asocial scientific education that the twentieth century favoured.

Being a member of an association confers an identity, which is useful in a profession where job security is low and freelances are plentiful. Science journalists' associations often offer press cards, literally identifying the bearer as a science journalist so that they can gain access to stories. They also offer services such as library access, which would otherwise be unaffordable to individual members (alongside the job insecurity, wages are low). It is typical of an association that it will be constituted as a charity or non-profit organisation, investing all income in the services for members. It thus sits outside the capitalist system.

By tracing the creation of science journalists' associations through time and across the world, this chapter considers the ways in which science journalists have organised in response to changing socio-economic conditions. This is not an exhaustive account: the formation of other organisations – such as Germany's Technical Literary Society, TELI, in 1929 (Goede, 2006) – can represent different alliances. Rather, the origins of the associations discussed here reveal key moments of geographically specific identity construction as science journalists have attempted to assert their distinctive contribution to the societies in which they work.

What triggers a science journalists' association? The case of the United States

Journalists have been forming groups to define their role for centuries. Studies of nineteenth-century journalism in the United States reveal debates about occupations, vocations and professions in a culture that was diverse, socially fluid and in search of identities and niches for new kinds of workers (Banning, 1999). Throughout the second half of the century, 'editorial associations' formed to consider ethics, standards and training. Sherrill et al. (2021) identify a continuous growth in the number of journalists' associations since the mid-1800s, but with a spurt in the early twentieth century, accompanying a huge expansion of press activity and the mass-marketisation of the daily press. There were also spikes after wars. A broadly similar pattern obtains in Northern Europe (Bensaude-Vincent, 1995; Le Cam, 2020).

As US journalism diversified into specialist beats, associations for specialist journalists also emerged. The National Association of Science Writers (NASW) was founded in 1934, with the stated motive to improve the craft and protect standards in science journalism (NASW, 2023). The move coincided with the dire economic conditions of the Great Depression, when science communication was being deployed both to promote the innovative high-technology projects of the New Deal policy and to encourage the roll-out of new technologies for existing industries, particularly in agriculture (Bevan and Smith, 2020).

Economic misfortune had provided an opportunity for science journalism, but for an association to flourish, there also needed to be the means for journalists to interact. The concentration of publishers in New York City and the pioneering use of the press conference (see, for example, Topper and Vincent, 2007) brought science journalists into close contact with their peers. An informal community therefore already existed that enabled the journalists to formalise their connections and provided a structure for welcoming others. This close community of science journalists – the 'Inner Club', as science communication scholar Sharon Dunwoody (1980) called it – has persisted. Dunwoody identified a more cooperative professional culture in US science journalism than in other fields where competition for stories, sources and readerships was the more usual professional relationship (see also Marcotte and Sauvageau, 2006).

The NASW's founder-members included John J. O'Neill, William L. Laurence, Howard W. Blakeslee, Gobind Behari Lal and David Dietz, who became the first president. Several of the group were migrants with strong political interests (Laurence was born Leib Wolf Siew and was involved in the ineffective Russian revolution of 1905). According to Lal, the journalists' ambitions for their beat were hampered by the reluctance of scientists to talk to them for fear of the disapproval of their peers and by a lack of respect from newspaper editors (Riess, 1983). The NASW, its founders hoped, would stabilise the profession, ensuring that their journalism would be trusted by scientists and editors and valued by readers. Membership at this time was highly selective and was promoted as a mark of quality and reliability. The association thus defined an elite who could win the confidence and trust of scientists and editors. The subsequent careers of this founding group – which includes several Pulitzer Prize winners – suggest that the strategy was successful.

Defining the role of "scientific journalism": The British science writers

The NASW strived to establish a particular identity for the science journalist in the space between the laboratory and the printing press, and in a particular policy context. In the UK, a similar process was in train, initiated by J.G. Crowther, who is now credited as Britain's first science journalist (Gregory, 2006). Crowther had been thinking about what could be done to improve the social relations of science for about two decades by the time he became a founding member of the Association of British Science Writers (ABSW) in 1947. In 1928, he had been hired as the *Manchester Guardian*'s 'Scientific Correspondent'. In an anthology of his articles published earlier that same year, he had set out the rationale, as he saw it, for science journalism: "the public should know more about science. Some knowledge of what scientists are doing, or hope to do, ought to be easily available" (Crowther, 1928: p. 35). Crowther believed that the application of science was fundamentally important to building a modern society and yet most people knew little science, especially those educated in the elite schools whose academic achievement consisted of "the study of the ancient classics" (Crowther, 1970: p. 43).

Popular science was changing at this time. Mass-market newspapers and the extension of democracy had made the public sphere a fractious and more potent arena, and, as in the United States, scientists who wrote for the public faced disapproval from their colleagues (Crowther, 1970; Gregory, 2005; Werskey, 1988). This environment demanded professional skills, but Crowther felt that the news journalists simply did not possess the necessary knowledge of science needed for science reporting. What he wanted, then, was "scientific journalists" who were both knowledgeable about science and writing "from the side of science" (Crowther, 1970: p. 48). Crowther later described himself as 'with', though not 'in', science.

In his next book, Crowther continued his campaign for 'scientific journalism':

> scientific journalism ... is becoming an essential binder in the structure of the civilisation [...]. Modern society may collapse unless the atmosphere of science becomes generally apprehended. [...] The proper scientific journalist should devote the whole of his intellectual energy to the prosecution of his craft.
>
> (Crowther, 1931: p. xiii)

Thus, the occasional contribution from scientists or other journalists was not enough: only dedicated 'scientific journalists' could prevent the collapse of civilisation. Many scientists were sympathetic to Crowther's approach and sought his acquaintance.

During the 1930s, Crowther travelled widely: in Nazi Germany, he helped persecuted scientists to emigrate; and in the United States in 1937, he met members of the newly formed NASW. The war would intervene, but Crowther was inspired to become a founding member of the ABSW a decade later. Among the other founding members – described by historian James Farry (2011) as a "self-appointed elite" – were experienced journalists Bill Dick, Arthur Woods Haslett, Anthony Michaelis and Maurice Goldsmith. Their politics were on the left: both Crowther and Goldsmith were suspected of spying for the Communists (Security Service, 1955; 1929-1949).

The ABSW would, its founders hoped, establish "science writing as a definite profession with appropriate status" (Goldsmith, 1949: p. 4), and in the early years, members included their ABSW membership in their journalistic by-lines. Membership was open to anyone who made a living from science writing, Crowther having conceded his 'scientific journalist' label for the broader 'writer' in the Association's title. *Nature* would later describe the early ABSW as "a kind of club for a somewhat embattled few", but acknowledged that it had soon flourished (Nature, 1969).

At one of its first meetings in 1947, the ABSW committed to easing "the tensions and prejudices between the norms of journalism and the scientific community", with the aim of protecting the scientists' values above those of their editors (Farry, 2011: p. 99). However, journalistic values were not to be discarded and members could be critical of science, as in 1953 when ABSW chairman Ritchie Calder made headlines after telling a meeting of the British Association for the Advancement of Science that "most scientists are illiterate, inarticulate, and irresponsible" (see, for example, Bradford Observer, 1953).

The association as an intermediary in society: The French experience

According to Anderson (1971), an association serves to link the individual to larger-scale institutions in society. In the UK, Crowther had negotiated a position between science and the media, and had declared himself with science but not in it. In France, the precise locations along these axes, and the loyalties they implied, were highly contested (Bensaude-Vincent, 1995; Bergeron, 2020).

After World War II, French bureaucrats reclaimed the palace in the salubrious 16th arrondissement that had housed the Nazi regime in Paris. UNESCO took offices there and in 1950 hosted the launch of the Association of Science Writers of France (l'Association des Écrivains Scientifiques de France, AESF). The first president was the educator and writer François Le Lionnais: a former chemical engineer and entrepreneur, he had fought for the Resistance and after the war became a director of studies at the Ecole Militaire.

Like Crowther, he argued that society had entered the "age of science", in which science journalism could provide life-long education for the people and facilitate a reasoned approach to the social challenges that would inevitably ensue (Le Lionnais, 1951).

At the AESF's launch meeting, the platform party included government ministers and the director-general of UNESCO, alongside Le Lionnais and Nobel laureate physicist Louis de Broglie. De Broglie stated that scientists should focus on research, rather than on popularisation, but worried that if scientists step away from public communication, their place might be taken by journalists who have little respect for accuracy, or who might, for commercial gain, "lower the level of their presentation in order to appear more accessible". The alternative, physicist and UNESCO officer Pierre Auger proposed, was science writers who were knowledgeable, responsive, authoritative and persuasive, and would create "a climate of enthusiasm and confidence" (Le Lionnais, 1951). In summary, the new Association was to facilitate the inevitable transformations of society by science, by supporting people with the right combination of scientific expertise and communication skills to generate public support for science.

The scientists, politicians and bureaucrats at the launch of the AESF were part of the national project to reorganise French science into a civil service. Coordination and direction was to be imposed from the top and science writers should perform their role as required. There were long-established niches for popular science in France and science writing was already a business and a profession, but science news journalists were nevertheless struggling for space in the national press. Newspaper editors expected science news to be written by scientists and endorsed by Academies. The stories were not integrated with current affairs and often appeared as small space-fillers at the bottom of the page. Another problem was the reluctance of scientists to talk to journalists, whose interest was considered intrusive. As in the UK, scientists who did talk to journalists were chastised for self-promotion.

In response to this situation, in 1955, four years after the launch of the AESF, a group of 16 science journalists formed their own Association of Science Journalists (AJSPI), which, according to one of them, Robert Clarke (n.d.), would "promote and defend" science journalism as a distinct category. They held their inaugural meeting not in a palace with a parade of dignitaries, but in the back room of a bar in the Opera district. On the political left, they, like their British colleagues, have been the subject of speculation about links to the Soviet Union (Wolton, 1993). The first president was physics-trained André Labarthe, the editor-in-chief of the magazine *Constellation*, which offered 'a French look at the world' – a concept developed by Labarthe while in war-time exile in London and explained there to an audience that included J.G. Crowther (Crowther, 1970). In a single edition from this time (April 1956), *Constellation* offered readers a food page, world news and the services of an 'agony aunt', as well as articles about uranium, cancer, cosmetic surgery, bird migration, Isaac Newton and national investment in atomic power.

Medical reporting was also a contested field at this time, with French journalists viewing medicine as a branch of science and so not distinct as a topic for reporting. Tensions had crystallised in 1952 when the news broke, on Christmas day, of the first kidney transplant (Fabiani, 2012; Piccoli et al., 2018). Marius Renard, an apprentice roofer who had ruptured his one good kidney in a fall, had received a transplant from his mother. Journalists reported Renard's slow death over the following three weeks, sometimes using unethical methods to source photographs and quotes from hospital staff. The national outpouring of grief exacerbated the hospital's discomfort at their very public failure. The medical authorities sued the newspapers and affirmed their non-cooperation with journalists.

The AJSPI campaigned for medicine, and any other branch of science, to be reported in the same way as crime or politics: by independent observers, and not by doctors, magistrates or ministers. AJSPI members emphasised the need for trust between reporters and the medical establishment. Journalists who specialised in medical reporting were happy under the 'science' umbrella for some years and only formed their own association in 1970 (Marchetti, 2017).

Further tensions surfaced when the French began testing atomic bombs in 1960. Journalists were not allowed to attend the test site. For the military, the tests were 'top secret', and the science journalists who reported on them found themselves in trouble. During the political upheavals of 1968, the AJSPI lobbied the government for the right to report on science, including such events as bomb tests. The politics of the situation was asserted by recent Nobel laureate Jacques Monod in 1969, when he stated that:

> It is up to the scientist to inform, but it is up to journalists to transfer this information to the public [...] it is their responsibility, it is important, and we must respect it, because it plays a part in democracy by involving citizens in decision-making.
>
> (Clarke, n.d.)

The AJSPI, then, showed some agility in managing its loyalties: it protected its members and at the same time identified a relationship with the scientific establishment, with the news industry and with government, adapting the balance to meet the needs of the moment.

Transnational associations: Standing together in Latin America

The US, UK and France established similar associations for similar reasons. As nations, however, they are far from typical of the rest of the world. Industrialised, wealthy and stable, they are democracies that promote a free press (at least in principle) and have long-established scientific institutions to keep the stories coming. The Anglophone North Atlantic region, in particular, continues to dominate global science news, putting science journalists in other countries in the role of language translator or international correspondents as the stories they cover are more likely to be from 'abroad' than from home (see Chapter 6).

The French AJSPI has members who are based overseas but report in French, sending stories from around the world to French news outlets. Francophone science journalism is, however, tiny in comparison to the global scale of Spanish-language reporting. Spanish-speakers are the second-largest language community on the planet (after Mandarin) and the most widely dispersed, most notably across South America. The long tradition of science scholarship and public communication in Spain provided a resource for popular science in its former colonies, often long before they had professional scientists of their own (Massarani, 2014). The easy communication with their Portuguese-speaking neighbours further broadened the community. Thus, the huge expanse of Latin America provided an extended market for science journalism, which grew rapidly in the late nineteenth and early twentieth centuries.

In 1951, the Argentinians, at the prompting of the mathematician, historian and editor José Babini, were the first on the continent to form a national association of science writers, which held its inaugural meeting within the Argentinian Association for the Advancement of Science (Calvo Hernando, 2005). It aimed to increase the amount and range of sciences communicated to the public; to ensure that science news was presented seriously and accurately; and to support and reward all the professionals in the science news process, including journalists, scientific advisors and translators.

In 1962, the Chileans hosted the first Inter-American Seminar on Science Journalism to be held in Latin America (for a detailed account, see Massarani, 2021). The continent was changing rapidly, through political experiments, cross-border trade agreements, improvements in public health and education, industrial innovation and an optimism that has been described as utopian (see, for example, Sorensen, 2007). Newspaper publisher John Reitemeyer, who was vice president of the Inter-American Press Association, lamented in his opening address that brilliant scientists in Latin America were often under-recognised and poorly paid, so they left to work overseas. He recommended that science and the media join together in a campaign to increase the public understanding of science and thereby generate greater support for scientists in their home countries.

The other speakers came from across the Americas, from either science or journalism. From further afield was Ritchie Calder of the ABSW, who had recently been awarded UNESCO's prestigious Kalinga Prize for the popularisation of science. The attendees comprised a broader group of scientists, science journalists (from Peru, Ecuador, Costa Rica, Argentina and Mexico) and some big funders: the Ford Foundation, the Rockefeller Foundation and the World Bank, which provided a focus on the role of science journalism in economic development.

Arguments from the podium were consistent with those of colleagues in the United States and Europe. Jacobo Brailovsky, who had written a science column for Argentina's *La Nacion* since 1924, addressed one recurrent question: is science journalism a job for scientists or journalists? Brailovsky argued that both can do it, but it is usually done better by journalists, as they have the necessary skills; he would go on to set up another Argentinian association, this one specifically for science journalists, in 1969. Venezuelan scientist, bureaucrat and populariser Marcel Roche – a future Kalinga Prize winner – argued that cooperation between journalists and scientists could help science to grow. He identified an ethos of service among science journalists in Venezuela who were committed to supporting the national research programme.

From Brazil came journalist José Reis, another future Kalinga winner. An accomplished virologist with international standing, Reis also wrote a column for the newspaper *Folha de S. Paulo* and had very recently been appointed its editor-in-chief. The paper exclusively reported the meeting on its front pages. Reiss argued that popular science could encourage positive attitudes towards science and so support the application of science to problems in society. Particularly where the educational system was weak, good-quality science journalism could compensate. Reis encouraged journalists to seek out stories at home, rather than overseas, to avoid overstating the power and potential of science and scientists. In Brazil, the government-sponsored José Reis Science Communication Award remains a high honour.

The 1962 meeting was the first of many. Similar events ensued in Ecuador, Argentina, Colombia and Madrid (Amorim et al., 2012; Calvo Hernando, 2005; 2006; Massarani, 2021). The trans-Atlantic link with Spain was formalised at the Madrid meeting in 1969 with the creation of the Ibero-American Association of Scientific Journalism, as had been proposed in the report of the 1962 meeting. This new transnational association was tasked with hosting a congress series and providing a platform for science journalists to develop further national associations.

The Ibero-American and national associations supported science journalism in Latin America at the local, national and international levels. However, the function of associations to provide linkages and mobility between different strata in society is severely impaired if those strata are themselves unstable. An association can only go so far in mimicking the values and practices of the hegemony if these are incompatible with the core values of the association. The turbulent politics of Latin America have thus presented a sometimes insurmountable challenge to science journalists' associations (Høyer, 2010; Waisbord, 2013; Sherrill et al., 2021). Argentina and Brazil have both suffered in this way. These countries' journalists sought shelter in international associations and have since been able to relaunch national associations in new formats.

Global science news

The twentieth century was rich in international associations: it produced, among others, the United Nations, the World Health Organisation, UNESCO and the World Bank. Three science-based developments intensified this global perspective: the planetary vision resulting from the space programme; the recognition in the 1960s of the environment as holistic and in crisis; and the world-wide deployment of digital technologies in the 1980s. These were not only global scientific events, but impacted too on global politics and economics.

The Ibero-American Association was not the first network of science journalists to have reached beyond national borders. The International Science Writers Association (ISWA) had first assembled two years earlier, in 1967, to support reporters who travelled to the United States to cover the space launches. With often limited English and little access to telephones, these reporters had previously relied on each other for informal support. The idea for a formal association came from a group of established journalists, including Gordon Rattray Taylor of the BBC, John Maddox of *Nature* and Dennis Flanagan of *Scientific American*. Rather than become an umbrella group for existing national associations, they opted for a membership of individual journalists. Although originally Anglophone and focused on the North-Atlantic region, the ISWA also served emerging nations, which valued its mentoring schemes and one-to-one teaching. At its peak, the ISWA had 400 members in 40 nations.

Joining the ISWA is one way for an individual to become a member of the World Federation of Science Journalists (WFSJ). Formed in 2002, the WFSJ is a federation of associations of science journalists, rather than of individuals. Nevertheless, it provides an opportunity for members of national associations to meet, network and gain support. It also provides an umbrella to shelter smaller associations. A good example is the case of Montenegro, which formed its national association of science journalists in 2016, catching up with its neighbours in Serbia and Croatia. Montenegro also joined the Balkan Association of Science Writers – a network that emerged from a regional seminar at the global science communication conference, PCST, in Italy in 2012, and which was encouraged by UNESCO's 2013 stimulus package for science journalism in the Balkans (Tatalović, 2013). In 2017, the Association of Science Journalists of Montenegro joined the WFSJ – a step that the Montenegrin chair Maria Bolevich hoped would improve the quality of science journalism in the country and reduce the amount of "copy-paste translations and pseudoscience" that she saw in the press (Tatalović, 2017). While newspaper editors in Montenegro had little interest in science coverage, membership of the WFSJ would, Bolevich hoped, provide "excellent motivation for all of us who want to do science journalism." This pattern of regional, sometimes language-specific, associations has been repeated across the world and on a variety of scales, as the complex interplay of connections and flow of information has been intensified by digital media.

Digital networks

When the Argentinian and Brazilian science journalists regrouped, it was not as 'associations', but as 'redes': networks (Massarani, 2021). In Argentina in 2007, 11 journalists began meeting up informally in a café and they quickly formed an online national network. Unlike the pioneers of the twentieth century, these 11 journalists were mostly women. Now with one hundred members, the Red Argentina de Periodismo Científico is concerned with supporting professional development for science journalists, engaging in critical reflection on science and society, and maintaining strong links between Argentinian science journalists and their colleagues around the world (RAPC, 2023).

In Brazil, the new RedeComCiência (Network of Science Journalists and Communicators) also participates in the WFSJ. This began in 2017 as a Facebook group and quickly gathered a wider membership. It soon faced the political and journalistic challenge of the Covid-19 pandemic, which was particularly acute in Brazil (Lancet, 2020). Very quickly, it established WhatsApp groups to support journalists on all beats who were now reporting on science and health in a sea of conflicting and controversial claims about the disease, and amid turbulent and sometimes dangerous national politics (De Ambrosio, 2020; Wendt, 2020).

Argentina and Brazil show how digital resources such as Facebook and WhatsApp can provide a platform for science journalists' associations. They operate on a wide range of scales and provide a toolkit that is emergent, mobile, flexible, accessible and cheap. The influence of these means for association is seen most clearly in the new science journalists associations across Africa. With the establishment of a digital infrastructure and access via smartphone, associations emerged across the continent, using wireless platforms such as Wordpress and LinkedIn to display the activities and values of science journalists within and beyond the community thus created. Across the wide diversity of peoples, languages, traditions, political systems and economic conditions, new social media provide a flexible, if flimsy, infrastructure that is open to all.

Digital networks can be easily made – and are easily broken. This fragility can become a survival tactic on a turbulent continent, where loyalties need to be continually under review and regrouping under a new flag needs to be relatively straightforward. The networks have the advantage over the associations of the twentieth century in that they operate on a variety of scales: one person with an internet connection can be associating with their neighbours, their colleagues across the continent and with others across the planet at one and the same time.

The Great Depression had been an impetus for the US's National Association of Science Writers and post-war reconstruction a context for European nations. Economic crisis and war are, however, chronic rather than acute conditions in Africa. Nor is a viral pandemic a novelty for countries still living with widespread HIV infection. In Africa, economic reconstruction, across a wide spectrum of often post-colonial conditions, is, as in Latin America in the 1960s, often labelled 'development'. Growth in science journalism in Africa has been encouraged, where resources permit, by the recognition that it supports the economic goals of the continent's policy emphasis on innovation (Lublinski et al., 2014). The aims of the new science journalists' associations in Africa reflect this context: while they seek to support a profession, the associations foreground their nations' intersected challenges of extreme poverty, poor public health and environmental decline, and ask what science might do to help (Nakkazi, 2012).

Kenya offers a good example of the complex challenges facing science journalists in Africa. Since declaring independence from the British Empire in 1963, Kenya has endured various forms of home-grown oppression, including a poorly regulated financial culture. It lost many people to AIDS in the late twentieth century, making it now (in 2023) a country of young people, many of whom are benefitting from improvements in basic public health infrastructure but who have limited access to education and employment. Regulation of the media has involved detailed oversight by the government and harsh financial penalties for its critics. This presents a challenging and fluid context for journalists of all kinds looking to make a career. According to a study of the structural factors constraining Kenyan journalism, journalists have had to deal not only with political interference, but also threats and physical attacks (Lohner et al., 2016; see Chapter 7 for similar pressures on Russian journalists). Each journalist has to decide how to operate in such a system, where the lack of job security can make them vulnerable to both financial inducements and self-censorship when they decide which stories to report.

The science base in Kenya is relatively strong for the region, with established universities and a constitutional commitment to economic development through innovation. The 'brain drain' has slowed, as Kenyan scientists are now more likely to find careers at home. Science R&D is driven by funding not from government, but through the old colonial links to Europe and investment from American corporations, whose focus on renewable energy and information technology (FCO, 2016) is rather distant from the priorities of ordinary Kenyans. Kenyan science is conducted in English, which, although a minority language in Kenya, fares better on the international stage than the Francophone journalism of the African countries further north. Yet, in looking outwards for its science, Kenyan science struggles to embrace the indigenous knowledge that sustains the environment and agriculture in the region and is intrinsic to local culture.

The Kenyan Environment and Science Journalists Association (KENSJA) was founded in 2004, at a time of political optimism and strong investment from overseas. It declares itself non-political and looks to inform decision-making about science and the environment by providing accurate and reliable information. It also provides a hub for professional development and local and international networking. While these activities mirror those of associations in the North, KENSJA also functions as a press office and knowledge mobiliser, particularly around agriculture, environment, health and development issues, and is accessible to many kinds of stakeholder. It campaigns for funding for media activity by scientific institutes and for open access to science publishing. Despite its non-political identity, in 2023, the lively KENSJA website, kensja.net, carried articles about US-funded fellowships for women policymakers to prioritise women's issues in agriculture, and an account of human rights abuses by renewable energy companies in their appropriation of land for wind farms. It is therefore less engaged with the national government, and more with international corporations, than any of its predecessor associations.

Kenya is also an active partner in the Nairobi-based transnational organisation Media for Environment, Science, Health and Agriculture (MESHA), which supports science journalists and encourages networking. According to MESHA's senior officer, science journalist Aghan Daniel (2023), there is a need for multiple associations in the region, and at a range of scales, because, given the opportunity presented by the innovation and development agenda, the task of science journalism is so large and diverse. Former executive director of the WFSJ, Jean-Marc Fleury, identifies that a "critical mass" of membership is needed for an association to thrive (Nakkazi, 2012). This can be achieved by transnational associations when national associations are small.

To the west of Kenya, Rwanda tells a different story. Emergent digital networks are responsive but also fragile, and, as associations are intermediaries, they require at least some stability in adjacent institutions if they are to survive. The Rwandan Association of Science Journalists (RASJ) was founded online in 2007 – a year with a sense of new beginnings for the country when, after a long period of internal conflict, Rwanda joined the East African trading bloc with support from the World Bank and the International Monetary Fund. The RASJ was set up by a small group of health and environmental reporters from different media institutions in Rwanda who wanted to raise awareness of the media's role in creating dialogue between scientists, policymakers and the public. They aimed to support and encourage training and career development for Rwandan science journalists and to promote "reliable science reporting" to support policy-making around science and technology research, particularly where it was relevant to life in Rwanda (RASJ, 2011).

The RASJ joined both the WFSJ and the short-lived but energetic African Federation of Science Journalists, which also encouraged science reporting with particular relevance to development. In 2008, the RASJ organised Rwanda's first conference on science communication and by 2011, it was looking forward to moving into new offices. However, the optimism did not last: this ambitious organisation, in a small, traumatised country, disbanded shortly thereafter. It had been a decade ahead of the institutionalisation of science in Rwanda, which established an Academy of Sciences in 2016 and a National Council for Science and Technology the following year. The Rwandan experience perhaps suggests that the science journalists' association – especially if the membership is small – cannot work on its own: it must be positioned among other institutions. Without infrastructure for science, policy and media, the association does not thrive.

Science communication has been an agent of colonisation since the eighteenth century, when it became an essential tool in the translation and reproduction of science and technology in places far from the laboratory (Alder, 1998). The same could be said of innovation policy today, which is an ideological and financial motor for many kinds of science journalism in many countries. Extending multinational and global associations of science journalists into new territories, as in Africa, risks imposing Western values and subjecting these regions to a normative model that may not be productive. According to Gervais Mbarga, a Cameroonian science journalist now teaching in Canada, "... one can fear that [foreign institutions] impose their way of seeing things, and not the way Africans see them." Mbarga notes that training for science journalists on African science or in local languages is rare, and "very little is said about what our African people think ... I hope it is not another strategy for killing African thought" (Nakkazi, 2012). For some African scholars, who see their world not as post-colonial but neo-imperial, the welcoming of African science journalists into international networks could be seen as a new form of oppression. The task for science journalists in Africa, according to sociologist Artwell Nhemachena (2020), is not just to grasp the helping hand of Western science, but also – and importantly – to elucidate and share the ideas about nature that have arisen in Africa from the world's oldest human societies. In theory at least, the same networks can carry all these messages.

Conclusion

Science journalists' associations and networks draw on a left-wing ideal that science, if openly discussed and thoughtfully applied, can make the world a better place. When the world shows clear potential for being better, such as during a depression or after a war, such associations find an opportunity. At the same time, these associations benefit individual journalists, who gain identity, training and rewards from their membership. Each participant will set their own balance of personal and societal gains from their membership, and this may change across a career, with more senior members mentoring and subsidising their junior colleagues. Membership across the span of a career may be more critical to the 'critical mass' than mere numbers.

Within the opportunity provided by large-scale political events, science journalists' associations draw on existing infrastructure: they stand on the shoulders of a strong news industry, a firm science base and a stable government. While they rely on this sturdy foundation, and can fail in its absence, the associations do not necessarily serve these institutions. Indeed, they are likely to be critical of them. Science journalists want more open science, better conditions of employment and the freedom to express their views. Where the individual is the primary unit and the Left is weak, as in the neoliberal environment, associations may struggle to provide a platform for negotiation among these institutions. The middle ground between the individual and the powers-that-be – be they state or corporate – may be too wide.

Nevertheless, professions need to show themselves to be useful, as this gives them their value to society. Science journalists' associations, like other professional associations, serve to highlight the beat's usefulness and keep their members trained, paid and celebrated. At the same time, they cannot always protect their members from scrutiny and exploitation. Indeed, they may draw unwelcome attention. Disassociating can therefore sometimes be a wise tactic, especially if a subsequent or alternative re-grouping is easy to achieve: 'weak links' are an important asset in any network. This raises the question, however, of what counts as an association in the digital age, if it can so easily be broken apart.

Through the transformation of the global economy in the late twentieth century, responsibility for the state of the world falls more and more to corporations rather than nation states. Corporations have different priorities from nation states and exercise different forms of power. Governments of all colours have policies for science and conventions and regulations for dealing with journalists – and, in any nation, these tend to be public knowledge. But corporations undertake research and deploy their press officers in mysterious ways that have thus far not been scrutinised by academics, at least not in science-based businesses. In some of the older science journalists' associations, press officers enjoy only a marginal membership. Closer – or perhaps simply more open – collaboration between science journalists and corporate press officers may be needed if the world of corporate science is to be scrutinised democratically. It is worth saying, too, that associations – sitting, as they do, outside of the capitalist system – are finding themselves occupying a dwindling sphere as the extension of capitalist logic throughout human activity continues apace. Perhaps other kinds of organisations will emerge under new conditions.

Corporations tend to extend across national boundaries and to carry values and practices from one part of the world to another, not unlike earlier forms of colonisation. What looks like investment in innovation in one place and welcome development in another can feel like subjugation or exploitation for some. The myth of science as universal has historically had its positive uses in diplomacy and the healing of conflict, but it is out of step with a world that is looking to the rich resource offered by human, cultural and epistemic diversity as the way forward. While the scientific establishment in the Global North confronts its own exclusivity and self-interest (Nature, 2022), science journalists from other regions can replenish the moral void with authentic visions of a survivable future. In a digital age, when individuals can own a smartphone and no one is more than six steps away from anyone they want to reach, it may not be necessary to associate in order to do that.

References

Alder, K., 1998. Making things the same: Representation, tolerance and the end of the Ancien Régime in France. *Social Studies of Science*, 28(4), pp. 499–545.

Amorim, L., Massarani, L., Bauer, M.W. and de Oca, A.M., 2012. Periodismo científico: Reflexiones sobre la práctica en américa Latina. *Chasqui. Revista Latinoamericana De Comunicación*, 120, pp. 73–77.

Anderson, R.T., 1971. Voluntary associations in history. *American Anthropologist*, 73(1), pp. 209–222.

Banning, S.A., 1999. The professionalization of journalism: A nineteenth-century beginning. *Journalism History*, 24(4), pp. 157–163.

Bensaude-Vincent, B., 1995. A public for science. The rapid growth of popularization in nineteenth century France. *Réseaux. Communication - Technologie - Société*, 3(1), pp. 75–92.

Bergeron, A., 2020. 'The Republic needs scholars!' A rapid history of making science public in 20th-century France. In: Gascoigne, T., et al. (eds.), *Communicating Science: A Global Perspective*. Canberra: ANU Press, pp. 297–316.

Bevan, B. and Smith, B., 2020. Science communication in the USA: It's complicated. In: Gascoigne, T., et al. (eds.), *Communicating Science: A Global Perspective*. Canberra: ANU Press, pp. 959–982.

Bradford Observer, 1953. British Association meetings: "most scientists are illiterate and inarticulate. *Bradford Observer*, 8 Sep., p. 3.

Calhoun, C. (ed.), 1992. *Habermas and the Public Sphere*. Cambridge: MIT Press.

Calvo Hernando, M., 2005. Ciencia y periodismo científico en Iberoamérica. *La Insignia*, Apr., https://www.lainsignia.org/2005/abril/dial_005.htm.

Calvo Hernando, M., 2006. *Arte y ciencia de divulgar el conocimiento*. Quito: Ciespal.

Clarke, R., n.d. L'association: Notre Histoire. AJSPI, https://www.ajspi.com/lassociation.

Crowther, J.G., 1928. *Science for You*. London: Routledge.

Crowther, J.G., 1931. *An Outline of the Universe*. London: Kegan Paul, Trench & Trubner.
Crowther, J.G., 1970. *Fifty Years with Science*. London: Barrie & Jenkins.
Daniel, A., 2023. Personal communication.
De Ambrosio, M., 2020. El valor del periodismo científico en tiempos de COVID-19. *SciDev.Net*, 15 May, https://www.scidev.net/america-latina/scidev-net-at-large/el-valor-del-periodismo-cientifico-en-tiempos-de-covid-19.
Dunwoody, S., 1980. The science writing inner club: A communication link between science and the lay public. *Science, Technology, & Human Values*, 5(30), pp. 14–22.
Fabiani, J., 2012. *Ces histoires insolites qui ont fait la médecine: Les transplantations*, Volume 2. Paris: Place des éditeurs.
Farry, J., 2011. *'Far more to it than appears on the surface': An historical investigation of the interface between space science and the British mass media*. PhD thesis, University of Manchester.
FCO, 2016. *UK Science & Innovation Network Country Snapshot: Kenya*. Foreign and Commonwealth Office, https://assets.publishing.service.gov.uk/government/uploads/system/uploads/attachment_data/file/564062/Kenya_Country_Snapshot_-_On_Template_-_30.9.16.pdf.
Goede, W.C., 2006. The twenties – exciting times in Germany. In: Dillsa, B. (ed.), *The Barriers Are Down: Eusja Advances Across Europe*. Turku: Enostone, pp. 9–17. https://www.teli.de/wp-content/uploads/pdf/eusja-book-teli-hr.pdf.
Goldsmith, M., 1949. Associations of science writers: The case for their stimulation and development. UNESCO. NS/PSI/2/REV; https://unesdoc.unesco.org.
Gregory, J., 2004. Who is the 'man in the street'? *RSA Journal*, 151(5512), pp. 28–31.
Gregory, J., 2005. *Fred Hoyle's Universe*. Oxford: Oxford University Press.
Gregory, J., 2006. Crowther, James Gerald (1899–1983), science journalist and administrator. *Dictionary of National Biography*. Oxford: Oxford University Press. doi.org/10.1093/ref:odnb/94975.
Høyer, S., 2010. The double bind of politics and profession. *Media History*, 16(3), pp. 267–280.
Lancet, 2020. Editorial: COVID-19 in Brazil: "So what? *Lancet*, 395(10235), p. 1461.
Le Cam, F., 2020. Journalistic organizations: Arenas for professional and symbolic struggles. *Oxford Encyclopedia of Communication*. Oxford: Oxford University Press. https://doi.org/10.1093/acrefore/9780190228613.013.831.
Le Lionnais, 1951. L'Association des écrivains scientifiques de France tient à l'UNESCO sa première séance publique. *Le Courrier De l'UNESCO*, IV(Mar), p. 9.
Lohner, J., Banjac, S. and Neverla, I., 2016. *Mapping Structural Conditions of Journalism in Kenya*. Media: Conflict and Democratisation, University of Leeds.
Lublinski, J., et al., 2014. Advances in African and Arab science journalism: Capacity building and new newsroom structures through digital peer-to-peer support. *Ecquid Novi: African Journalism Studies*, 35(2), pp. 4–22.
Marchetti, D., 2017. Information under control: Medical news in France between the early 1950s and early 1980s. *Australian Journalism Review*, 39(2), pp. 37–49.
Marcotte, P. and Sauvageau, F., 2006. Les journalistes scientifiques: Des éducateurs? *Les Cahiers du Journalisme*, 15(Hiver), pp. 174–195.
Massarani, L., 2014. Science communication in Latin America: What is going on? *Science Museum Group Journal*. https://dx.doi.org/10.15180/140205/001.

Massarani, L., 2021. Jornalismo científico na américa Latina: Registro histórico do primeiro seminário interamericano realizado na região em 1962. *Intercom: Revista Brasileira De Ciências Da Comunicação*, 44(1). https://doi.org/10.1590/1809-58442021113.

Nakkazi, E., 2012. The rise of African science journalism. *SciDev.Net*, 19 Oct., https://www.scidev.net/global/features/the-rise-of-african-science-journalism-1.

NASW, 2023. About the National Association of Science Writers Inc., https://www.nasw.org/about-national-association-science-writers-inc.

Nature, 1969. Obituary: A.W. Haslett. *Nature*, 223(2), p. 545.

Nature, 2022. Collection: Diversity, equity and inclusion in science. *Nature*, 17 Jan., https://www.nature.com/collections/dafichiff.

Nhemachena, A., 2020. *Securitising Monstrous Bottoms in the Age of Posthuman Carnivalesque?: Decolonising the Environment, Human Beings and African Heritages*. Langaa: Bamenda.

Osborne, T., 2021. Civil society, populism and liberalism. *International Journal of Politics, Culture, and Society*, 34(2), pp. 175–190.

Piccoli, G.B., Richiero, G. and Jaar, B.G., 2018. The pioneers of nephrology – Professor Gabriel Richet: "I will maintain". *BMC Nephrology*, 19, p. 60. https://doi.org/10.1186/s12882-018-0862-0.

RAPC, 2023. Orígenes de la Red. Red Argentina de Periodismo Científico, https://radpc.org/quienes-somos/origenes-de-la-red.

RASJ, 2011. Rwanda Association of Science Journalists, 9 Jun., https://www.comminit.com/content/rwanda-association-science-journalists-rasj-1.

Riess, S., 1983. A Journalist from India, at Home in the World: Oral History Transcript; Interview with Gobind Behari Lal. Bancroft Library, Regional Oral History Office, University of Berkeley.

Security Service, 1929-1949. Communists and suspected communists. James Gerald Crowther, KV 2/3341. The National Archives, London.

Security Service, 1955. Communists and suspected communists. Maurice Joseph Goldsmith, KV 2/4048. The National Archives, London.

Sherrill, L., et al., 2021. Journalism's backstage players: The development of journalism professional associations and their roles in a troubled field. *Journal of Media Business Studies*, 19(1), pp. 53–71.

Sorensen, D., 2007. *A Turbulent Decade Remembered: Scenes from the Latin American Sixties*. Stanford: Stanford University Press.

Tatalović, M., 2017. Association of Science Journalists of Montenegro accepted for membership of the World Federation of Science Journalists. *Balkan Network of Science Journalists*, 15 Jan., https://balkansciencejournalists.wordpress.com/2017/01/15/association-of-science-journalists-of-montenegro-accepted-for-membership-of-the-world-federation-of-science-journalists.

Tatalović, M., 2013. Ministers in South-East Europe urged to support science journalism. *EuroScientist*, 15 Mar., https://www.euroscientist.com/ministers-in-south-east-europe-urged-to-support-science-journalism.

Topper, D. and Vincent, D., 2007. Einstein's 1934 two-blackboard derivation of energy-mass equivalence. *American Journal of Physics*, 75(11), pp. 978–983.

Waisbord, S., 2013. *Reinventing Professionalism: Journalism and News in Global Perspective*. Oxford: John Wiley & Sons.

Wendt, L.G., 2020. RedeComCiência lança iniciativa para combater desinformação acerca do coronavírus. *RedeComCiencia*, 5 May, https://www.redecomciencia.org/post/redecomci%C3%AAncia-lan%C3%A7a-iniciativa-para-combater-desinforma%C3%A7%C3%A3o-acerca-do-coronav%C3%ADrus.
Werskey, G., 1988. *The Visible College: Scientists and Socialists in the 1930s*, 2nd ed. London: Free Association Books.
WFSJ, 2023. World Federation of Science Journalists, wfsj.org.
Wolton, T., 1993. *Le Grand Recrutement*. Paris: B. Grasset.

PART II

Journalist-source relations

4

COPY AND PASTE

Churnalism in science journalism

Lars Guenther, Justin T. Schröder and Anna Tratter

You've had a long and stressful day at work. After finally returning home, you decide to catch up on the latest news as you usually do. You turn on the television for the evening news bulletin. At the same time, you start browsing through online newspapers and checking the latest social media posts on your mobile phone. You are particularly interested in science, so when the television news presenter starts talking about a new development in nanotechnology, you pay attention. But wait! Haven't you just read on your phone the exact same words the TV presenter is now saying? And isn't that picture in the background of the TV studio oddly familiar?

If you have ever experienced something like this, then the chances are you were encountering churnalism. Also known as copy-and-paste journalism, this occurs when journalists rely heavily on external third-party material – including news agency copy, press releases from non-media organisations, reports in other news outlets and social media posts – and then use it with little verification or editorial input, resulting in stories that are either reproduced verbatim from the external source or consist of varying combinations of editorial and third-party material. There is some evidence that the prevalence of churnalism has increased in recent years, and it appears to be especially widespread in science journalism. This chapter examines why the science beat may be particularly prone to churnalism.

Churnalism is not a new phenomenon. As early as the 1970s, studies identified that a significant proportion of journalistic output could be traced to public relations (PR) activity (Bentele and Nothhaft, 2008). These early studies characterised the relationship as a one-way influence in which public relations operations determined the topics, timing and issue selection of news journalism. Later scholars argued that the relationship was more complex than

DOI: 10.4324/9781003326724-7

this determination model implies. For instance, the intereffication model, proposed by Günter Bentele and colleagues in 1997, described the ways in which both journalism and PR adapt in response to the needs and routines of the other, in a system of mutual influences and dependencies (Bentele and Nothhaft, 2008; see also Comfort et al., 2022; Machill et al., 2006). Even so, journalists' dependency on external material is frequently judged to be undermining professional standards, as the term 'churnalism' suggests. Bahtiyar Kurambayev is blunt, referring to the practice as plagiarism; Juliane Lischka and colleagues designate it "obsequious" journalism (Kurambayev, 2020; Lischka et al., 2016).

Some researchers use a narrow definition of churnalism as the convergence between journalism and PR specifically; in other words, as part of the 'PR-isation' of journalism (Berglez, 2018; Heyl et al., 2020; Jackson and Moloney, 2016). However, others use a broader definition encompassing all types of external, third-party material or 'subsidised content' (Boumans, 2017: p. 2264; see also Boumans et al., 2018; Lewis et al., 2008; Skärlund, 2020; Van Leuven et al., 2014; Wheatley and O'Sullivan, 2017). In this chapter, we follow the broader definition, which allows interdependences to be considered. For instance, news agency copy, coverage of other media, or social media posts could all themselves be influenced by PR material. PR may therefore indirectly affect journalism even when these other types of external material are used. News agencies, in particular, have undergone many changes in recent decades and seem susceptible to a heightened reliance on external material (Boumans et al., 2018; Lewis et al., 2008; Lorenz, 2017). Journalists may therefore be influenced by PR material without even being aware of it. In addition, digital platforms have brought with them different attitudes towards third-party material. Most explicitly, online news aggregators are founded on the premise that it is acceptable to select and re-circulate other outlets' news coverage (Johnston and Forde, 2017; Skärlund, 2020).

In what follows, we consider the prevalence, causes and implications of churnalism in the reporting of science news. We place the case of science journalism in the context of journalism more generally, examining the pressures on the profession as a whole as well as those characteristics of the science beat that leave science journalists especially exposed to PR capture.

The prevalence of churnalism

Evidence for the prevalence of churnalism across journalism as a whole is mixed (Van Hout and van Leuven, 2017). Some studies conclude that it is widespread in a number of countries and regions – for instance, in the UK (Davies, 2009; Lewis et al., 2008), Sweden (Almgren, 2017; Skärlund, 2020), Central Asia (Kurambayev, 2020) and Zimbabwe (Mabweazara, 2013) – and in a variety of beats, including cultural affairs (Kristensen, 2018), crime

(Ndhlovu, 2020) and celebrity journalism (Van den Bulck et al., 2017). After surveying over 150 studies from a century of research into the relationship between journalism and PR, public communication scholar Jim Macnamara concludes that "50% to 75% of mass media content is provided or significantly influenced by PR" (2016: p. 118). There is also some evidence that its presence is increasing; for instance, Saridou and colleagues (2017) found that churnalism in Greece had increased between 2013 and 2016. However, other researchers have found little evidence for widespread churnalism, including in some of the same countries (such as the UK) where others had found it to be common (see Nicholls, 2019, for the UK; Boumans, 2017, for the Netherlands; Lischka et al., 2016, for Switzerland; and Van Leuven et al., 2013; 2014, for Flanders). The discrepancies in these findings could arise from different criteria for what is considered to be 'journalistic' reporting or from different types of news outlets being sampled. There is some indication that quality news outlets rely less on external materials (Boumans, 2017; Boumans et al., 2018) than do popular outlets; that shorter items are more dependent on external material than longer pieces (Lewis et al., 2008); and that timing is also an important factor, with churnalism being more likely when there is less time between the press release and the news report (Schafraad and Van Zoonen, 2020). Churnalism is also more likely to be found in online news than in print-first publications (Wheatley and O'Sullivan, 2017).

The high rate of churnalism in science journalism seems more clear cut (Schäfer, 2017; Williams and Gajevic, 2013). In the case of the UK, an analysis of the BBC's coverage of science found that about three quarters of its news about scientific research relied on press releases, with a further eighth of such items deriving from other forms of PR activity or appearing to do so (Mellor et al., 2011: p. 4). This study found that the press releases had typically been issued by scientific journals, universities or research organisations. Similarly, a study tracing the news coverage of a scientific publication about a possible link between meat consumption and pancreatic cancer found that three sources (the press release from the *British Journal of Cancer*, a BBC story and a story on the National Health Service's website) accounted for more than 85% of the news items (Taylor et al., 2015). Many of these articles were verbatim copies, or only slightly edited, and included no credit to the copied source. Another UK study provides evidence that science journalism and related beats might be especially vulnerable to churnalism compared to other beats. Justin Lewis and colleagues compared the coverage of different topics in national news outlets. They did not include a separate category for science, but they found that health news – a topic that makes up a considerable proportion of many science journalists' output – was associated with the most PR-generated material (Lewis et al., 2008).

In other countries, too, science journalists churn external materials, with one study finding that up to 65% of the sentences used in science articles in

US news outlets have a high similarity to press releases (Comfort et al., 2022). However, this study found differences depending on the reputation of the media organisation. Those with a history of award-winning science reporting relied less on the PR material. Similarly, a study of South African science journalism found that news outlets published press releases from universities with minimal input from their own journalists and no crediting of the original source (Heyl et al., 2020). Health journalists working for magazines in Belgium have been found often to lack transparency about the use of PR as a source, although when interviewed, the journalists agreed that the impact of PR and advertisers on their work was increasing (De Dobbelaer et al., 2017). It is not only PR material that is churned. A US study found that although health reporters rely on university press releases, they make even greater use of materials sourced from other media (Len-Rios et al., 2009).

In science reporting, as in general journalism, the prevalence of churned news depends on the type of outlet, with churnalism being more common in online outlets than in legacy news outlets – as we show below with an example of chemistry news coverage. Ownership models may also be a factor. For instance, in German television coverage, commercial stations are more likely than public stations to broadcast footage from video news releases by scientific organisations and the use of such footage results in more frequent and longer coverage (Machill et al., 2006). The newsworthiness and credibility of the sources of PR are also likely factors influencing whether it will be reproduced with minimal alteration. A Dutch study found that university press releases with stronger news values are more likely to be published unchanged (Kroon and Schafraad, 2013; for more mixed findings, see Schafraad and Van Zoonen, 2020).

In a global survey of science journalists, there was general agreement that science PR was driving science news and that this resulted in a lower quality of science journalism (Massarani et al., 2021). Some science journalists, however, deny engaging in churnalism (McKinnon et al., 2018). This is not necessarily inconsistent with the evidence that science churnalism is commonplace, as it could result when general news journalists are assigned to science stories. However, the studies mentioned above do also identify examples of science specialists churning PR copy (e.g., Mellor et al., 2011; Williams and Gajevic, 2013).

A short example tracing the use of PR copy helps illustrate the nature of churnalism within science journalism. The German Chemical Society (Gesellschaft Deutscher Chemiker, GDCh) was founded in 1949 to represent the interests of German chemists regionally, nationally and internationally, and today has a membership of about 30,000.[1] The Society maintains its own press service (Pressedienst Chemie), which, among other things, is responsible for issuing press releases. In 2020, the Society issued 18 press releases, 17 of which attracted media coverage, mostly in online outlets but also in print.[2]

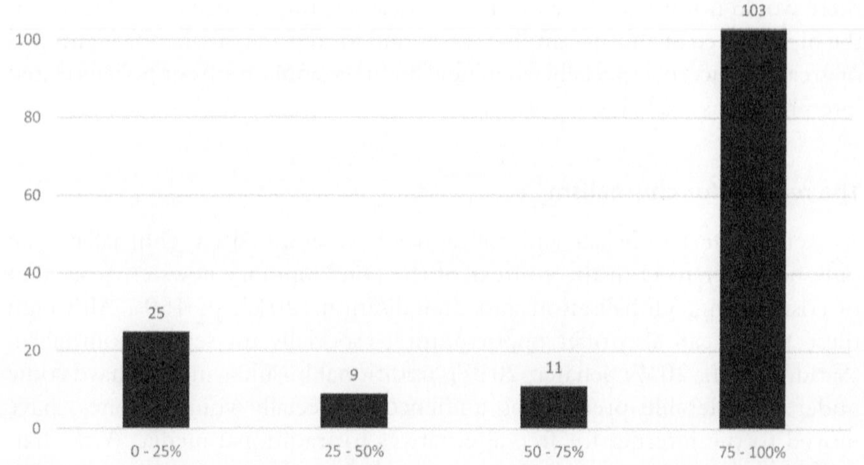

FIGURE 4.1 Degree of similarity between German Chemical Society press releases and subsequent journalistic reporting.

We were able to retrieve 148 of the subsequent journalism articles, which we compared with the original press releases to assess the proportion of the text that was identical to the press release. We found that the average similarity score for the articles was 75% (see Figure 4.1); that is, three quarters of the text in the average article was copied directly from the press release. Only about a sixth of the articles had a similarity score less than 25%. Furthermore, 58 of the articles (39%) were published verbatim. Some of these had been shortened compared to the press release, but 42 (28%) articles were identical to the press release in all respects. Thus, almost a third of the articles had no alteration to the press release whatsoever. Three quarters of the articles attributed the source, but this same statistic also means that a quarter of the articles failed to mention the source.

As noted above, there were differences in the extent of the churnalism depending on whether the article appeared in print or online only. Online articles had significantly higher similarity scores (83% on average) than did print articles (just 33% on average)[3] and not a single print article was identical to the press release. There were also differences across the topics covered. The highest degree of churnalism was found in the reporting of GDCh events, which on average had similarity scores of 85%, followed by articles about prizes and awards with average similarity scores of 84%, chemical/scientific topics with similarity scores of 76% and, lastly, offers by the GDCh with just 40% similarity. For announcements of events and prizes or awards, it may not be of great concern that science journalists use information verbatim, especially when it comes from a reputable scientific organisation such as the GDCh;

there would not usually be a need for critical scrutiny in such cases. However, the high degree of churnalism that was found for reports of scientific matters is of greater concern, especially when new findings, applications or policy-related interpretations are being reported.

The reasons for churnalism

As Rebeca De Dobbelaer and colleagues have emphasised, churnalism can only be understood in the context of the contemporary news environment of cost-cutting, globalisation and digitalisation (2017: p. 450). Although these trends can also offer opportunities, especially for science journalism (Saridou et al., 2017; Schäfer, 2017), traditional business models have come under considerable pressure as audiences, especially younger ones, have moved to the internet for free alternatives to traditional media. With that, advertisers have also moved. News companies' attempts to recover costs with online editions have, more often than not, failed to make up for the loss in subscriptions and advertising revenue (Mitchell et al., 2016). Financial pressures experienced by some media companies have led to their closure. At the outlets that have survived, falling revenue has often affected the viability of specialised beats. This has included science journalism (Bauer et al., 2013; De Dobbelaer et al., 2017; Schäfer, 2017), a beat that even in good times probably never generated much income (Brumfiel, 2009). Many editors perceive other news beats to be more important, cheaper and having a wider appeal, which at the same time also make them more attractive to advertisers. Science news is therefore often seen as an expendable luxury (Allan, 2011; Schäfer, 2017).

Much has been written about the decline of science journalism, especially in Western countries, including diminishing job security, cuts to staffing and salaries and precarious working conditions for those journalists who remain (Allan, 2011; Bauer et al., 2013; Brumfiel, 2009; Bubela et al., 2009; Heyl et al., 2020; Massarani et al., 2021; Schäfer, 2017; Smith and Morgoch, 2020; Williams and Clifford, 2009; Williams and Gajevic, 2013). As is true of journalism in general, the remaining journalists rely on fewer resources and experience higher workloads. As a result, they lack the time to spend on their reporting, including checking facts or finding additional sources through independent investigation. Additionally, when science desks are closed and generalists take over the science coverage, the expertise to carry out these tasks may be missing. In both cases, the reliance on external material may increase.

These trends are all the more pressing in online environments that follow faster 24/7 news cycles. Online news is subject to stronger competition between media companies to fill news holes with fewer resources and to be the first to report on current developments (Boumans, 2017; Boumans et al., 2018; Rulyova and Westley, 2017). Real-time user metrics mean that journalists and their editors now focus more on audiences' preferences in order to compete in the online market for attention. Journalists of all types, including those on the science beat, are now required to report on multiple platforms – for instance, articles, podcasts and social media (Brumfiel, 2009; Jackson and Moloney, 2015; Johnston and Forde, 2017). Overall, then, fewer professionals must produce more content in less time (De Dobbelaer et al., 2017), a combination that makes churnalism an attractive – and perhaps the only – option (Eliott et al., 2018; Jackson and Moloney, 2015). While some science journalists complain that working their way through the hundreds of emails they receive from PR organisations each day only increases their workload (see Williams and Clifford, 2009), external material that arrives in the form of page-ready copy allows news organisations to produce low-cost and spreadable news, to remain competitive and to provide constant news updates. This stands in contrast to the expense and long timescales required for original investigative reporting.

What are existential pressures for news organisations can be seen as opportunities for organisations seeking PR opportunities to secure media coverage. As PR material is promotional in nature, intended to boost an organisation's image and enhance its reputation, in most cases it will not be designed to align with the public interest remit of journalism. Yet the existence of press material alone is a powerful predictor of media coverage (Comfort et al., 2022; Taylor et al., 2015). As effective public relations operations require significant financial resources, elite organisations and actors who can afford to engage in PR activity are able to gain privileged access to the news media (De Dobbelaer et al., 2017; Van Leuven et al., 2014).

Scientific bodies are among the elite organisations that invest in public relations. Science PR has been increasing since the 1970s, not only in response to the changes in the media environment but also due to additional pressures on scientists to publicise their work (Bauer et al., 2013; Weingart and Guenther, 2016). Research organisations, especially universities, increasingly encourage scientists to engage in communication to meet demands for public accountability, build institutional profiles and reputations and compete for resources – a situation that is also encouraged by major research funders. This trend can also be linked to the medialisation of science, the process identified by sociologist Peter Weingart (1998) whereby scientists and their institutions have become re-oriented towards the mass media as the latter becomes an ever more important resource for the public legitimacy of science (see also Comfort et al., 2022; Heyl et al., 2020; Weingart, 2017; Williams and Gajevic, 2013).

The medialisation of science leads research institutions to put significant effort into gaining media attention and generating positive news coverage, at the same time the vacuum created by the changes in the media sector is increasing the likelihood that these PR efforts will succeed. For instance, Daniel Vogler and Mike Schäfer (2020) find that an increasing amount of Swiss news coverage is influenced by university press releases and includes positive evaluations of the university. In line with new policy directives for external visibility and public engagement, increasingly scientists also communicate directly with public audiences themselves; through blogs, posts on social media, or articles on digital-native outlets, like *The Conversation*, aimed at giving academics and other experts a public voice. Consequently, says science journalist Geoff Brumfiel, writing at a time when many scientists were first taking up blogging, scientists and their institutions have "more influence than ever over what the public reads about their work" (Brumfiel, 2009: p. 275).

As the sources of potential science news ramp up their public communication efforts, the nature of the science beat may also increase their chances of securing media coverage. Science journalists are highly dependent on their sources due to the specialised, and often esoteric, expertise the sources provide. While scientists' readiness to communicate with public audiences can make the journalist's work easier (Bubela et al., 2009; Guenther and Joubert, 2021; Schäfer, 2017), science journalists' reliance on expert sources leaves them vulnerable in the face of expanding PR operations. Some of the journalists Brumfiel spoke to in 2009 feared that the combined pressures of science PR, newsroom cuts and scientists' own public communication were leading to the end of the science beat within traditional newspaper journalism. The worst of these fears have not been realised. In 2021, a global survey of science journalists found that 73% of respondents did not agree with the statement that "science journalism is dying out" (Massarani et al., 2021). Nevertheless, many of the journalists in the survey agreed that the beat is in crisis; only a third disagreed, down by 12% compared to a similar survey carried out eight years before. Thus, the pressures on science journalists continue, and possibly increase, contributing to the conditions in which churnalism becomes an accepted practice.

The implications of churnalism

Churnalism has significant implications for the nature and quality of journalism, both for journalism in general and for the science beat in particular. On a structural level, many researchers believe that it weakens both the profession and the quality of what is produced, particularly in regard to its potential contribution to the proper functioning of democracy (Almgren, 2017; Lorenz, 2017). Churnalism reduces the likelihood that reporting will be independent in presenting a balanced variety of topics, actors and frames or holding to account those in influential positions – including in science. Media scholar Peter Berglez (2018) fears that intensified links between sources (he focuses on politicians), journalists and PR practitioners violate the unique duties and roles that each of them has within a democratic society. Over-reliance on external material and using it almost verbatim means the material is being re-produced uncritically. This distorts the values and norms that journalists are known for, such as their autonomy or their fourth estate function of providing democratic accountability. It also means less journalism-driven, investigative reporting.

The journalistic norm of balance (see Chapter 8) is also undermined. The analysis of the BBC's science coverage mentioned above found that stories based on PR material often lacked other sources (Mellor et al., 2011). An over-reliance on external material thus leads to single-source journalism in which no additional, independent voices are heard (Guenther, 2019). Balance is lacking not only due to this absence, but also because the agenda and framing of the story have been set by the providers of the external material. Thus, as researchers Daniel Jackson and Kevin Moloney have put it, external material is not only an ***information*** subsidy but also an "***editorial*** subsidy" (2015: p. 769). Furthermore, over-reliance on external material leads to duplication of the same material in different outlets across a range of platforms. As portrayed in the scenario at the beginning of this chapter, audiences are repeatedly exposed to the same story and most likely to the same framing. Journalists, especially those answerable to a centralised news desk, experience pressure to run stories out of fear that others will cover them first (Williams and Clifford, 2009). Thus, churnalism affects news circulation, journalistic originality, quality and media diversity, resulting in a homogeneity across news outlets akin to pack journalism (Dwyer and Martin, 2017; Lorenz, 2017; Nicholls, 2019).

On an individual level, churnalism also affects journalists' working conditions and their conceptions of their role. These include the transformation of journalistic work from outside news gathering into desk-bound work and challenges to their roles as gatekeepers, processors of information or information brokers (Boumans, 2017; Boumans et al., 2018; Davies, 2009; Jackson and Moloney, 2015; Lewis et al., 2008; Van Hout and van Leuven, 2017). Many science journalists, in particular, fear that with increasing churnalism, traditional – and especially critical and investigative – roles are being lost. If not journalists, who will hold scientists to account as spenders of public money and influencers of public policy? While there are global differences, most science journalists see themselves as neutral informers and translators of scientific information (Guenther, 2019; Massarani et al., 2021). As Luisa Massarani and colleagues noted in their report on global science journalism, the essential contribution of investigative and critical science reporting becomes most apparent in media environments where freedom of expression is curtailed (see also chapter 7); such conditions make clear that quality information is "not only an economic asset but also a public responsibility" (Bauer et al., 2013:

p. 5). The risk is that in accepting the agenda-setting and framing of news by news actors, churnalism reduces science journalism to celebratory stories about the achievements of science; and that in so doing, it recasts science journalists as cheerleaders rather than as critical watchdogs who interrogate scientific claims in order to safeguard the public interest.

Conclusion

While the death of science journalism may have been exaggerated, the science beat – like journalism generally – is transforming in response to significant pressures. As we have discussed, churnalism is widespread in science news and, although most studies have focused on Western countries, it appears to be a global phenomenon. Discussions of churnalism in general news reporting typically link it to the high workloads that are symptomatic of crisis conditions. Tom Van Hout and Sarah Van Leuven (2017) identify two distinct responses to this situation: one sees churnalism as an existential threat; the other, more optimistic view, sees it as a feature of journalistic practice in transition. The first position regards churnalism as fundamentally undermining the profession of journalism. For instance, Bahtiyar Kurambayev (2020), examining the consequences of churnalism in Central Asia, argues that the widespread use of ready-to-print external materials actively encourages plagiarism, creating a journalistic culture that values verbatim copying and sending misleading signals to early-career journalists that plagiarism is an accepted journalistic norm. By this account, churnalism is upending journalistic standards for a future generation of reporters. As we have discussed, the central issue here is the way in which a dependency on materials generated by news actors may hinder journalism's democratic and public service functions, something that is also of concern even to those working in PR (Jackson and Moloney, 2015). This focus on the news media's role as the fourth estate highlights the loss of science journalists' independence as they surrender their role as critics and watchdogs.

To Suzanna Comfort and her co-authors (2022), this is a matter of who holds power: journalists or their sources? By enabling public relations operations to set the news agenda and frame how stories are reported, churnalism has the potential to increase the influence of those who already wield significant power within society. Scientists and their employing organisations may not think of themselves as powerful, but their epistemic authority is vital to the exercise of power in the knowledge society. Science churnalism can promote the interests of influential scientific journals that are owned by multinational corporations, or of elite universities operating in a global market, or of individual scientists seeking to promote a particular policy position for ideological reasons. Even when it promotes publicly funded disinterested research, science PR shares the aims common to all forms of public relations: enhancing visibility, attracting funding, building reputations and advertising a product (in this case, usually a research paper). Scientific press releases are written with the interests of the organisation at heart. Indeed, in extreme cases, it is the writers of the press releases, not the journalists, who generate the hype found in the subsequent news reports (Allan, 2011; Heyl et al., 2020). The medialisation of science means that the constant pressure for PR opportunities also infects the production of science itself, so that it can become difficult to identify where the science ends and the PR begins.

Murcott and Williams (2013) suggest that this situation not only does journalism a disservice, but science too. The lack of critical reporting and the reproduction of promotional materials could result in an unrealistic picture of research that overpromises with claims of unambiguous breakthroughs and unquestioned certainties, when in reality, science is tentative, incomplete and hard to interpret. Since the credibility of science is dependent on the credibility of its communication, public trust in science is likely to be impacted when unwarranted promises fail to deliver. If churnalism becomes widely recognised by public audiences, it could therefore erode public trust in both journalism and science (King, 2019; Macnamara, 2016; Weingart and Guenther, 2016).

However, arguments about the ill effects of churnalism are, as Van Hout and Van Leuven note, premised on "the reassuring but lofty idea that journalism safeguards democracies from corruption, falsehood, and malgovernance" (2017: p. 118). This idealistic vision of a pre-PR era fails to address how PR-free journalism could survive under current conditions. An alternative response to churnalism in general news journalism is to see it as a necessary adaptation as the news media transition to the digital environment. From this perspective, there is nothing inherently wrong with journalism that relies on external material. Rather, what matters is where the external material comes from and how it is used.

Indeed, if an organisation is an advocate of the public interest, its PR activity may even be welcomed. For instance, looking at the Flemish press coverage of Medicine Sans Frontiers, Sarah Van Leuven and colleagues (2013) argue that press releases from NGOs function as a source of balance to the PR of more powerful actors. A similar case could be made for environmental NGOs or health advocacy groups, and also perhaps for research organisations since most of these are public bodies oriented towards a social good. For example, in a report commissioned by the BBC Trust, geneticist Steve Jones claims that many scientific press releases are "objective and straightforward" (BBC Trust, 2011: p. 40). Journalists can differentiate, this line of argument implies, between PR about publicly or philanthropically funded research on the one hand, and industry-funded research on the other. If the former is indeed "objective and straightforward", as Jones supposes, then an increased, but discriminating, reliance on science PR could lead to news reporting that is more accurate (from a scientific point of view). The same applies to some other types of external material. For instance, provided it is of good quality, the widespread use of news agency copy aligns with the aim of enabling a range of outlets to publish accurate and unbiased science news.

Nor does the extensive use of PR mean that journalists cease all journalistic activity. A proliferation of PR from multiple sources, including a variety of different types of organisations, requires journalists to continue to exercise gatekeeping functions, drawing on their professional values to identify which press releases are worth publishing and which should be binned. Journalists working on a press-released story must also make a distinction between what can be published verbatim (usually in shorter pieces) and stories they need to spend more time on and engage with more thoroughly. Nor is journalism just about writing original material; deleting information from a press release also requires journalistic judgement. Van Hout and van Leuven (2017) suggest that recognising these journalistic activities for what they are requires a better understanding of the digital environment where eyewitness reporting is no longer the most appropriate mode of operation.

External materials are time-saving resources that potentially free up journalists' time for background research, contextualisation and investigative journalism. If this is the goal, then the key question is whether there are alternative ways to support science journalists other than a dependence on promotional PR. This may mean alternative business models that can pay for traditional journalistic work, or it may mean rethinking what roles are necessary for science journalists to take on a watchdog role in the digital media environment. Among the new financing options are sponsored science news desks (Bauer et al., 2013; Bubela et al., 2009), crowdfunding (Schäfer, 2017) or enticements for consumers to go back to paying for quality journalism. Changing roles for science journalists have been explored in some Science Media Centres (see Chapter 5) or at outlets such as *The Conversation* where the journalists' role is as editor rather than writer (Guenther and Joubert, 2021). Increasing digitalisation may also open up new roles for science journalists, perhaps blending with the roles of other science communication professionals in encouraging public engagement with science through more dialogue and building public trust through greater transparency.

Regardless of how science journalists' roles evolve in response to the changing media environment, churnalism is likely to continue for as long as science PR meets the needs of both the scientific organisations that produce it and the news organisations that use it. For now, despite the potential threats it poses, churnalism helps science journalism survive in a commercial market that attaches little value to public-interest journalism.

Notes

1 https://www.gdch.de/.
2 The press releases and media clippings were gathered through personal contacts at the GDCh. The journalism items consisted of 24 print articles and 124 online items. 32% ($n = 47$) of the articles' topic concerned chemical or scientific topics; 33% ($n = 49$) prizes and awards; 16% ($n = 24$) offers by the GDCh; and 19% ($n = 28$) were about GDCh events. The Preposteo plagiarism software was used to compare the articles with the press releases.
3 $t(146) = -7.119, p < .001$.

References

Allan, S., 2011. Introduction: Science journalism in a digital age. *Journalism*, 12(7), pp. 771–777.

Almgren, S.M., 2017. Undoing churnalism? *Digital Journalism*, 5(8), pp. 1060–1079.

Bauer, M.W., et al., 2013. *Global Science Journalism Report: Working Conditions and Practices, Professional Ethos and Future Expectations.* London: SciDev.Net.

BBC Trust, 2011. *Review of Impartiality and Accuracy of the BBC's Coverage of Science.* London: BBC Trust. https://www.bbc.co.uk/bbctrust/our_work/editorial_standards/impartiality/science_impartiality.html.

Bentele, G. and Nothhaft, H., 2008. The intereffication model: Theoretical discussions and empirical research. In: Zerfass, A., van Ruler, B., Sriramesh, K. (eds.) *Public Relations Research*, Wiesbaden: Springer. 10.1007/978-3-531-90918-9_3.

Berglez, P., 2018. Smileys without borders: A critique of transboundary interaction between politicians, journalists and PR practitioners on social media. *Triple*, 16(1), pp. 18–34.

Boumans, J., 2017. Subsidizing the news? Organizational press releases' influence on news media's agenda and content. *Journalism Studies*, 19(20), pp. 1–19.

Boumans, J., Trilling, D., Vliegenthart, R. and Boomgaarden, H., 2018. The agency makes the (online) news world go round: The impact of news agency content on print and online news. *International Journal of Communication*, 12(2018), pp. 1768–1789.

Brumfiel, G., 2009. Supplanting the old media? *Nature*, 458(7236), pp. 274–278.

Bubela, T., et al., 2009. Science communication reconsidered. *Nature Biotechnology*, 27, pp. 514–518.

Comfort, S.E., Gruszczynski, M. and Browning, N., 2022. Building the science news agenda: The permeability of science journalism to public relations. *Journalism & Mass Communication Quarterly*, pp. 1–20.

Davies, N., 2009. *Flat Earth News.* London: Vintage.

De Dobbelaer, R., Van Leuven, S. and Raeymaeckers, K., 2017. Dirty dancing: Health journalists and the pharmaceutical industry a multi-method study on the impact of pharma PR on magazine health news. *Public Relations Review*, 43(2), pp. 450–459.

Dwyer, T. and Martin, F., 2017. Sharing news online. *Digital Journalism*, 5(8), pp. 1080–1100.

Eliott, J., et al., 2018. An examination of Australian newspaper coverage of the link between alcohol and cancer 2005 to 2013. *BMC Public Health*, 18(47), pp. 1–13.

Guenther, L., 2019. Science Journalism. *Oxford Research Encyclopedia of Communication.* 10.1093/acrefore/9780190228613.013.901.
Guenther, L. and Joubert, M., 2021. Novel interfaces in science communication: Comparing journalistic and social media uptake of articles published by *The Conversation Africa. Public Understanding of Science*, 30(8), p. 1041–1057.
Heyl, A., Joubert, M. and Guenther, L., 2020. Hype and churnalism in science communication – Comparing university press releases and journalistic articles in South Africa. *South African Journal for Communication Theory and Research*, 46(2), pp. 126–145.
Jackson, D. and Moloney, K., 2016. Inside churnalism. *Journalism Studies*, 17(6), pp. 763–780.
Johnston, J. and Forde, S., 2017. Churnalism. *Digital Journalism*, 5(8), pp. 943–946.
King, A., 2019. Polluted at the faucet. *EMBO Reports*, 20(12), e49600. 10.15252/embr.201949600.
Kristensen, N.N., 2018. Churnalism, cultural (inter)mediation and sourcing in cultural journalism. *Journalism Studies*, 19(14), pp. 2168–2186.
Kroon, A.C. and Schafraad, P., 2013. Copy-paste of journalistieke verdieping? Een onderzoek naar de manier waarop nieuwsfactoren in universitaire persberichten nieuwsselectie en redactionele bewerkingsprocessen beïnvloeden. *Tijdschrift Voor Communicatie Wetenschap*, 41(3), pp. 283–303.
Kurambayev, B., 2020. The causes and consequences of plagiarism by journalists in Central Asia. *Asian Studies Review*, 44(4), pp. 691–708.
Len-Rios, M.E., et al., 2009. Health news agenda building: Journalists' perception of the role of public relations. *Journalism & Mass Communication Quarterly*, 86(2), pp. 315–331.
Lewis, J., Williams, A. and Franklin, B., 2008. A compromised fourth estate? *Journalism Studies*, 9(1), pp. 1–20.
Lischka, J., Stressig, J. and Bünzli, F., 2016. News about newspaper advertisers: To what extent can corporate advertising budgets predict editorial uptake and coverage of corporate press releases. *Journalism*, 18(10), pp. 1397–1414.
Lorenz, H., 2017. News wholesalers as churnalists? *Digital Journalism*, 5(8), pp. 947–964.
Mabweazara, H.M., 2013. Normative dilemmas and issues for Zimbabwean print journalism in the "information society" era. *Digital Journalism*, 1(1), pp. 135–151.
Machill, M., Beiler, M. and Schmutz, J., 2006. The influence of video news releases on the topics reported in science journalism. *Journalism Studies*, 7(6), pp. 869–888.
Macnamara, J., 2016. The continuing convergence of journalism and PR: New insights for ethical practice from a three-country study of senior practitioners. *Journalism & Mass Communication Quarterly*, 93(1), pp. 118–141.
Massarani, L., Entradas, M., Fernandes Neves, L.F. and Bauer, M.W., 2021. *Global Science Journalism Report 2021. Working Conditions and Practices, Professional Ethos and Future Expectations.* London: SciDev.Net.
McKinnon, M., Howes, J., Leach, A. and Prokop, N., 2018. Perils and positives of science journalism in Australia. *Public Understanding of Science*, 27(5), pp. 562–577.
Mellor, F., Webster, S. and Bell, A.R., 2011. *Content Analysis of the BBC's Science Coverage.* London: BBC Trust. https://www.bbc.co.uk/bbctrust/our_work/editorial_standards/impartiality/science_impartiality.html.

Mitchell, A., Holcomb, J. and Weisel, R., 2016. *State of the News Media*. Washington, DC: Pew Research Center. http://www.journalism.org/2016/06/15/state-of-the-news-media-2016/.
Murcott, T.H.L. and Williams, A., 2013. The challenges for science journalism in the UK. *Progress in Physical Geography: Earth and Environment*, 37(2), pp. 152–160.
Ndhlovu, M.P., 2020. Examining media discourses on religious rape in Zimbabwe. *Feminist Media Studies*, 20(6), pp. 801–812.
Nicholls, T., 2019. Detecting textual reuse in news stories, at scale. *International Journal of Communication*, 13, pp. 4173–4197.
Rulyova, N. and Westley, H., 2017. Changing news genres as a result of global technological developments. *Digital Journalism*, 5(8), pp. 986–1005.
Saridou, T., Spyridou, L.-P. and Veglis, A., 2017. Churnalism on the rise? Assessing convergence effects on editorial practices. *Digital Journalism*, 5(8), pp. 1006–1024.
Schäfer, M., 2017. How changing media structures are affecting science news coverage. In: Jamieson, K.H., Kahan, D. and Scheufele, D.A., *The Oxford Handbook of the Science of Science Communication*. New York: Oxford University Press, pp. 51–60.
Schafraad, P. and Van Zoonen, W., 2020. Reconsidering churnalism: How news factors in corporate press releases influence how journalists treat these press releases after initial selection. *Communications*, 45(1), pp. 718–743.
Skärlund, S., 2020. The recycling of news in Swedish newspapers: Reused quotations and reports in articles about the crisis in the Swedish academy in 2018. *Nordicom Review*, 41(1), pp. 69–84.
Smith, H. and Morgoch, M.L., 2020. Science & journalism: Bridging the gaps through specialty training. *Journalism Practice*, 16(5), pp. 883–900.
Taylor, J.W., et al., 2015. When medical news comes from press releases—a case study of pancreatic cancer and processed meat. *PLoS ONE*, 10(6), e0127848. 10.1371/journal.pone.0127848.
Van den Bulck, H., Paulussen, S. and Bels, A., 2017. Celebrity news as hybrid journalism: An assessment of celebrity coverage in Flemish newspapers and magazines. *Journalism*, 18(1), pp. 44–63.
Van Hout, T. and van Leuven, S., 2017. Investigating 'churnalism' in real-time news. In: Franklin, B. and Eldridge, S. (eds.), *The Routledge Companion to Digital Journalism Studies*. New York: Routledge.
Van Leuven, S., Deprez, A. and Raeymaeckers, K., 2013. Increased news access for international NGOs? *Journalism Practice*, 7(4), pp. 430–445.
Van Leuven, S., Deprez, A. and Raeymaeckers, K., 2014. Towards more balanced news access? A study on the impact of cost-cutting and web 2.0 on the mediated public sphere. *Journalism*, 15(7), pp. 850–867.
Vogler, D. and Schäfer, M.S., 2020. Growing influence of university PR on science news coverage? A longitudinal automated content analysis of university media releases and newspaper coverage in Switzerland, 2003-2017. *International Journal of Communication*, 14, pp. 3143–3164.
Weingart, P., 1998. Science and the media. *Research Policy*, 27(8), pp. 869–879.
Weingart, P., 2017. Is there a hype problem in science? If so, how is it addressed? In: Jamieson, K., Kahan, D. and Scheufele, D. (eds.), *The Oxford Handbook of the Science of Science Communication*. New York: Oxford University Press, pp. 111–118.

Weingart, P. and Guenther, L., 2016. Science communication and the issue of trust. *Journal of Science Communication*, 15(5), pp. 1–11.
Wheatley, D. and O'Sullivan, J., 2017. Pressure to publish or saving for print? *Digital Journalism*, 5(8), pp. 965–985.
Williams, A. and Clifford, S., 2009. *Mapping the Field: Specialist Science News Journalism in the UK National media*. Cardiff: Cardiff University.
Williams, A. and Gajevic, S., 2013. Selling science? *Journalism Studies*, 14(4), pp. 507–522.

5

SCIENCE MEDIA CENTRES

Walking a line between science, PR and journalism

Irene Broer

It's early morning at a newsroom in central Cologne. One of the editors, Clara, greets me at the door with a mug of Milchkaffee in hand. Like other newsrooms across Germany, the open-plan office will soon be buzzing with the sound of typing and talking, but for now, it's just Clara and me, preparing for the morning editorial meeting where the staff will decide which topics to pursue for today's science news. Yet this is not a typical newsroom. Clara, who has a PhD in biology, is one of eight science editors at the Science Media Center Germany. Rather than produce content for their own public-facing outlet, Clara and her colleagues produce free-to-use science content for other journalists.

Since the first Science Media Centre (SMC) was established in the UK in 2002, similar centres have been set up around the world. The centres function as intermediary organisations, supporting the relationship between scientists and journalists by providing press briefings, summaries, expert statements and background information on controversial science topics and new research. SMCs have been celebrated for their ability to facilitate journalists' access to scientific expertise, serving as what sociologist Simone Rödder (2020) has described as an "organised contact system". However, they have also been criticised for setting the news agenda on scientific issues with public policy implications. A particular point of contention has been the extent to which SMCs, rather than assisting public interest journalism, facilitate 'science PR' on behalf of academic institutions, researchers and publishers. Taking the case of SMC Germany, this chapter examines the ways in which such intermediary activity constructs – and blurs – the boundaries between PR and journalism.

Journalism can itself be understood as having an intermediary function, mediating as it does between news actors and the public (Matthews, 2014).

DOI: 10.4324/9781003326724-8

However, the professional norms and expectations of journalism can suppress the flexibility that is required of intermediary organisations, which need to be able to adapt to the demands and goals of the actors they mediate between. In the case of science journalism, for over a century, there has been a perception that the news coverage of science could be improved through the involvement of additional intermediary organisations that are separate from the news outlets. As described in Chapter 1, in the United States, science journalism developed as a specialist newsroom beat in conjunction with the establishment of the Science Service news wire in the 1920s. Like today's SMCs, Science Service's aim was to mediate between scientists and news outlets by providing the latter with copy-ready science stories. Emphasising the truth-searching similarities between science and journalism, the founders of Science Service believed that "mobilising science" would simultaneously serve the interests of science and journalism, and through the latter, the general public (Ritter, 1926: p. 200).

Today, similar assumptions underpin the growing number of intermediaries that seek to influence the relationship between science and journalism. As well as SMCs, these include university press offices, science press release platforms such as EurekAlert! and journalism outlets like *The Conversation* and *SciDev.net* that are based on collaborations with academics (Guenther and Joubert, 2021). As also discussed in Chapter 4, each of these intermediaries can influence the prioritising, interpretation and presentation of scientific information in the news media, thereby shaping the way that scientific issues are framed in public debate.

Whether an intermediary organisation aligns itself with science or with journalism – or genuinely occupies neutral territory in between – depends in part on the circumstances of its foundation. SMC Germany, unlike other SMCs, was founded not by academic or political elites but by science journalists. This identification with journalism shapes SMC Germany's editorial practices and professional norms and values. It also informs its self-understanding as an independent non-profit organisation providing reliable material for journalists working in increasingly demanding and precarious circumstances. However, this model of not-for-profit journalism operating at one remove from public audiences faces similar constraints to regular newsrooms – including staff shortages, the demands of news rhythms and values and the need to access scientific sources – all of which affect the extent to which it can develop as a new form of journalism.

This chapter is based on ethnographic observations of SMC Germany carried out in January and October 2020. During these visits, I kept detailed field notes of editorial meetings, the team's pitching of stories and their production efforts. In addition, I had access to the Center's written documents and online office chat and I conducted interviews with all the members of the SMC Germany team.[1] In what follows, I draw on this material to examine how the daily routines and practices of the SMC Germany newsroom navigate the boundaries between journalism and PR. First, though, I consider how this intermediary space had previously been configured, by outlining the contrasting motivations behind the British and German SMCs.

A press office for science or a service for journalists?

SMCs have developed as a response to perceived problems with science journalism. The impetus behind the first SMC in the UK was a supposed "crisis of trust" in science that had been highlighted in an influential report published in 2000 by the House of Lords Science and Technology Select Committee. In the preceding decade, the UK had experienced a series of high-profile science-based controversies over the safety of BSE-infected beef, genetically modified organisms and the measles, mumps and rubella vaccine. Both scientists and politicians believed that poor-quality media coverage was implicated in what they took to be falling levels of public trust in science. Better-quality science reporting, the reasoning went, would mean better-informed citizens who would be more likely to support science and technology and to understand the implications of scientific and technological developments. The report concluded that the coverage of science stories by general news journalists was "the great challenge" and that this challenge could be met in two ways: "by changing the behaviour of the media, and by changing the behaviour of scientists in dealing with the media" (House of Lords, 2000: para.7.26).

The report is credited with prompting Baroness Susan Greenfield, then Director of the Royal Institution in London, to set up an advisory council to determine how to improve the reporting of science by general news journalists (SMC, 2002). By the end of 2001, a director and a media relations assistant had been appointed to a new 'Science Media Centre' and provided with offices at the Royal Institution, where they were to "work on behalf of scientists and the scientific community as a whole" (SMC, 2002: p. 13). The SMC would, Greenfield stated, be "unashamedly pro-science" while also "independent of any particular agenda" (p. 14). The centre was officially launched in 2002, supported by funding from scientific bodies and corporate donors. It was to focus on improving the quality of "science in the headlines" by providing non-specialist journalists with the information and support needed to produce accurate, balanced and in-depth coverage of science and technology issues.

In the words of its first – and two decades later still its only – director, Fiona Fox, the SMC is a "press office [...] for the whole of science" (Fox, 2012: p. 257). This close alignment with science immediately provoked criticism. Even before the new centre had opened its doors, public participation researcher Tom Wakeford (2001) referred to it as "Britain's first Ministry of Truth" and claimed it could "help bring about a new totalitarianism of knowledge." A decade later, science journalist Connie St Louis (2013) concluded that: "It would not be harsh to call the London-based centre a PR agency." A content analysis by media researchers Andy Williams and Slavko Gajevic (2013) found that the SMC had played a significant role in a media campaign in favour of research on animal-human hybrid embryos ahead of a vote in parliament on a controversial bill, an amendment to which had proposed banning such research. Scientists' messages dominated the subsequent media coverage and the amendment was rejected. Such influence, Williams and Gajevic suggest, raises concerns that SMC activities might not improve, but rather harm, both the quality of journalism and public trust in science.

Sociologist Joan Haran has argued that in emphasising, as she puts it, the "scientistic voice which attributes rationality and accuracy to the experts", the SMC ignored the inclusive, public-centred approach to public science that had been called for in the Lords' report (Haran, 2012: p. 245). Similarly, in a study commissioned as part of a review to explore whether there was a need for an SMC in Germany, Simone Rödder identifies the British SMC as a "technical fix": "Its deficit model-informed communication policy is that the political is technical, the technical should be evidence-based and this evidence should come from scientific experts" (Rödder, 2015: p. 387).

Despite such criticisms, SMCs have since become a global phenomenon. The UK initiative was followed by SMC Australia in 2005, New Zealand in 2008 and Canada in 2010, before spreading – albeit sometimes only temporarily – beyond the Anglosphere to Japan, Germany, Taiwan, Malaysia, Kenya, India and Spain. SMCs have also been considered in France, Denmark, Italy, Norway, Pakistan and Indonesia (Foucart and Horel, 2020; Scimex, 2014). A Global SMC Network held its first meeting in June 2011 (SMC, 2012: p. 36). Like the British organisation, the other SMCs also often work with universities, research institutes, government agencies and other organisations to connect journalists with subject experts and provide other sources of information. They may also offer training to help journalists improve their understanding of complex scientific concepts and issues. Though each operates independently, all SMCs in the global network follow the same mission as the UK original: "to inform public debate and discussion on the major issues of the day by injecting evidence-based science into headline news" (SMC Global, 2012).

While the original SMC in the UK was founded on the initiative of government and scientific elites, SMC Germany was initiated by members of the Wissenschaftspressekonferenz (WPK), the German professional association for science journalists. Rather than a decline in public trust of science, these journalists were primarily worried about the quality of science journalism at a time of budget cuts and fast-paced news flows. Developments in digital media have led traditional media outlets to experience losses in advertising revenue and subscriptions, leading to a decline in the number of specialised science journalists (Dunwoody, 2021; Fahy and Nisbet, 2011; Schäfer, 2017). The constant news cycle and the rise of social media have meant that science journalists, like other journalists, must produce more content more quickly, making it difficult to engage in in-depth reporting (Bauer et al., 2013; Massarani et al., 2021). At the same time, the proliferation of digital platforms provides scientists with the opportunity to share their work directly with public audiences, bypassing traditional gatekeepers such as science journalists. While this has given scientific actors more control over how their research is communicated, it has also raised concerns about the accuracy and objectivity of science communication in the absence of professional journalism, especially given the impact of algorithms on the dissemination and amplification of controversial views (Brossard and Scheufele, 2013). This not only impacts public audiences directly, but also further increases the burden on science journalists, who are faced with an oversaturation of information (St Louis, 2013). The decreases in both financial and knowledge resources for science reporting thus risk an increase in sensationalised and polarised coverage of scientific issues (Franzen, 2012).

Veteran science journalist Volker Stollorz and colleagues from the WPK saw potential for the SMC model to mitigate some of these pressures. In 2012, they initiated an "exploration phase" to investigate setting up such an organisation in Germany (Hettwer et al., 2013). This involved a survey of colleagues, a consultation with leading communication researchers and an invitation to the director of the UK's SMC to speak at a science journalism conference in Germany. While the SMC approach seemed a promising answer to the perceived difficulties in German science journalism, there was concern about the British SMC's self-proclaimed goal of promoting science in policy debates. The report of the exploration phase concluded that German science journalism could benefit from a SMC, but that it was imperative for the organisation to focus on addressing the needs of science journalists rather than scientists. To avoid being seen as performing science PR for particular interests, SMC Germany was to be a service *for* journalists, *by* journalists (Hettwer et al., 2013). The idea received backing from the Klaus Tschira Foundation (KTS) and the new organisation was launched in 2016. KTS continues to be the main funder, along with around 50 universities and research facilities, government ministries, news outlets, professional associations, foundations and private donors (SMC Germany, 2022).

In the newsroom: Curating science, creating stories

By the time of my visit to SMC Germany in early 2020, the newsroom had established a routine based on familiar journalistic practices. Clara starts her morning with an overview of the latest developments in science. This is no easy feat. Researchers in academia are under pressure to increase their visibility and publish frequently. Currently, about two and a half million scientific papers are published each year – around 6,800 a day – a number which seems to be on the rise (Bornmann and Mutz, 2015; Ware and Mabe, 2015). At the same time, science is becoming more specialised, leading to a proliferation of outlets for research findings, including niche journals and online preprint repositories (Gu and Blackmore, 2016; Taubert and Weingart, 2017). This fragmentation makes it difficult for scientists and non-scientists alike to keep up with the latest developments in scientific fields. As well as increasing numbers of academic publications, scientific practice is increasingly 'medialised', oriented towards media logics in order to gain public visibility (Peters, 2012; Rödder et al., 2012; Väliverronen, 2021; Weingart, 2001). Academic organisations, scientific journals and individual scientists all seek to increase their visibility through various means, such as professionalising their public relations departments, distributing press releases and promoting scientific results on social media (Franzen, 2014; Friedrichsmeier et al., 2015; Raupp, 2017). Visibility can pay off: studies have shown that scientists who use social media are more likely to have their research covered in the media and that academic publications promoted through press releases are more likely to be cited by other scientists (Lemke et al., 2022; Phillips et al., 1991; Priem et al., 2012).

Like other science journalists, Clara and her colleagues navigate this abundance of source material with the help of the embargoed press services of major scientific publishers, such as *Nature*, *Science*, *JAMA* and *PLOS*. These journals' newsletter-style emails list the titles of 50 to 100 research articles scheduled for publication and are sent out to journalists on a weekly basis to give them a head start on their reporting. They also provide a preselection of those studies that the journals consider most likely to be attractive to media outlets (Franzen, 2012, 2014; Kiernan, 2006). As with the science desks at other newsrooms (Blöbaum et al., 2017; Granado, 2011), journals are SMC Germany's main source of new scientific content and provide a crucial time advantage for SMC editors to create content before the embargo ends and media reports can be published.

After noting down some of the research articles to follow up on later, Clara moves on to the news media. International, national and local outlets can all reveal which stories are starting to attract significant public attention. An automatic news alert helps with this task, flagging up content related to SMC Germany's main scientific interests of health, medicine, climate, environment, artificial intelligence, energy, mobility and technology. This 'news radar' enables SMC Germany to keep track of public debate on scientific issues, the experts who are being quoted and the narratives that are circulating about these topics. This media monitoring can reveal scientific studies that Clara and her colleagues have hitherto overlooked or the beginnings of a science-related controversy that warrants further attention.

At 10 a.m., it is time for the morning editorial meeting. The editorial staff gather around Clara's desk, some on rolling chairs, others standing or leaning against the glass dividers, pens and paper at the ready. Clara reads from her list. She starts with the media coverage of previous SMC Germany stories, which the organisation tracks by searching the national media for titles, sentences and scientists' statements derived from SMC content. The editors debate why one topic has received more attention than another and comment on the use of their work. The expert quotes they gather are intended for journalists to use without crediting SMC Germany. However, consistent with the widespread churnalism discussed in Chapter 4, some media outlets are in the habit of also copying SMC Germany headlines, teasers and body texts without any adaptation. While this is not forbidden, SMC Germany content is intended as a building block for other journalists' reporting, with the summaries and expert statements provided by SMC Germany to be woven into larger stories. SMC output should be considered "raw material", explains the deputy editor-in-chief: "We're not looking to take away journalists' work, but we do want to make it easier by doing the prep work. A baker needs eggs, flour, milk – raw material – but what they make is entirely up to them. That's what we aim for here."

When selecting new topics to pursue, the editors consider how much media attention the stories should or might receive – or are already receiving. In the case of new studies, the results should not only be scientifically relevant but also considered a public issue that could potentially – and realistically – affect the lives of many citizens. A ground-breaking pharmaceutical intervention to treat Alzheimer's disease is unlikely to be selected if it has only been tested on mice. A meta-study on the global decline of insects, on the other hand, would certainly be selected. The SMC input in such cases might have an 'agenda setting' role. However, in a move that differs from traditional journalistic practice, the selection of some stories may have an 'agenda blocking' function if editors suspect that a scientific claim could easily be misinterpreted or is already being exaggerated or hyped in the media. In case of doubt, the deputy editor-in-chief will invariably remind all the staff to consider what their "colleagues out there" – the accredited journalists who have registered to receive SMC Germany's content – need most.

The transition from a topic into a story also requires the SMC to have access to the relevant scientific publications and to scientists willing to provide statements, in-depth knowledge of the issue, a journalistic 'gut feeling' to assess the scientific relevance and newsworthiness and, crucially, the time to research it (Broer and Pröschel, 2021). The stories that SMC Germany editors finally agree on are, on average, the result of an hour and a half of editorial discussions, often continued in the in-house chat channels.

After deciding on the stories they will be working on that day, the editors prepare to contact scientists for expert assessment ("Einschätzen lassen"). First, they summarise the topic and identify a newsworthy angle based on the reason it caught the editors' interest in the first place. Then, they formulate questions; for example, asking scientists to consider the novelty, quality and reliability of a new study, or to verify and contextualise claims being made in public debate. This leads the editors to the next decision: who is a suitable expert? SMC Germany has a large database of scientists, each listed with one or more tags describing their discipline and specialty. If none are found in the database, a search on Google can help. The most important criteria, according to Volker Stollorz, the SMC editor-in-chief, is that a scientist is part of the scientific community with a proven track record of research in the appropriate field, as judged, for instance, by having publications in high-ranking scientific journals.

Most of SMC Germany's output is text-based and, like the journals' embargoed emails that the SMC itself works from, is sent as emails to its accredited journalists before being published on the website some days later. In common with other SMCs, including the original British one, formats include 'research in context', 'rapid reaction', 'fact sheet' and 'science response' – each of which has different purposes and is aimed at different levels of urgency. 'Research in context', for instance, usually anticipates that an area of science is about to trigger controversy. Media researchers Andy Williams and Slavko Gajevic describe the use of these interventions by the British SMC as "pre-emptive attempts to manage news coverage" (2013: p. 513). By contrast, the 'rapid reactions' are more reactive and aimed at contextualising or managing issues that are already receiving media attention.

The emails always address the recipient journalists as "dear colleagues" ("Liebe Kolleginnen und Kollegen"). They start with a teaser formulating the angle of the story and some text summarising the issue at hand, followed by statements from expert scientists. These are organised by expert, each listed with their names, titles and institutions, and tend to be rather long by journalistic standards – 250 words per quote is not uncommon. The original questions posed by the SMC editors are typically not included in the text, except in the 'science response', which takes the form of a Q&A. The stories may be for immediate release ("Zur sofortigen Verwendung") or may be embargoed until a specific date and time (the "Sperrfrist").

An alternative format is the press briefing. Originally run as in-person events focusing on broad, longer-running topics, these underwent significant changes during the Covid-19 pandemic and are now run online (Broer, 2020). As well as providing journalists with information on current issues at the forefront of public debate, press conferences serve as a form of relationship management between journalists and scientists, allowing for direct, but streamlined, two-way communication (Broer and Pröschel, 2022; Rhein and Lautenschläger, 2022; Williams and Gajevic, 2013).

Despite SMC Germany's self-identification with journalism ("the colleagues out there") and its adoption of standard journalistic routines, the similarity of its products with those of PR is striking. In contrast to journalism's external representation of collective interests, in its most basic function PR is the self-representation of particular institutional interests (Baerns, 1985: p. 149). It can be understood as the interest-driven communication of an organisation with its environment: a "structural coupling", in the words of linguist and PR consultant Cathrin Christoph (2009: p. 82), in which a complexity reduction takes place for a journalistic audience (Hoffjann, 2007). Science PR, specifically, refers to the strategic use of communication by scientific organisations and institutions – for example, by organising press conferences, issuing press releases, providing background information to journalists and arranging interviews with scientists. While often conflated with other goals, science PR has the aim to promote organisations' research activities and increase their impact for their own benefit (Raupp, 2017; Weingart et al., 2021).

With its embargoed releases, provision of background information and occasional press conferences, the SMC Germany newsroom produces the same outputs as most PR offices. Furthermore, although its ready-to-use texts, statistics and quotes are intended as practical support for journalists, SMC Germany also seeks to influence the direction of public debate on science topics by setting the agenda and framing the narrative. Yet the organisation's interest-driven communication is geared in two directions, neither of them aimed at promoting the interests of individual science organisations. First, it seeks to support journalism by providing the knowledge and practical solutions that overworked and under-resourced journalists need to do their jobs (Broer, 2020; Buschow et al., 2022; Suhr et al., 2022); and second, it aims to raise the status of scientific knowledge and science-informed journalism in public debate. This became particularly clear during the Covid-19 pandemic.

A critical moment

Like many media organisations, SMC Germany experienced what journalism scholars Thorsten Quandt and Karin Wahl-Jorgensen (2021) have called a "critical moment" during the pandemic, as its editors had to grapple with both the scientific uncertainty and implications of the novel coronavirus, on the one hand, and extreme time pressure to deliver information to journalists, on the other. Yet the pandemic also gave the organisation an opportunity to re-define its mission and role in science communication.

As the virus started to spread, the organisation adapted and extended its editorial practices. For example, in addition to the usual outputs, it began to provide journalists with contextualised statistics ("statistische Aufklärung") and rapid assessments of pre-print publications, rather than waiting for these to first pass through peer review and appear in the scientific journals. At the same time, the in-house research-and-development team, SMC Lab, developed free-to-use software to assist journalists in their reporting. In the following months, SMC Germany began to take on a new role as an "expert on expertise", including sharing guidelines on how to determine scientists' credibility and how to deal with unverified data in preprints. SMC Germany's efforts did not go unnoticed in the journalistic community: in 2020, its audience grew from approximately 800 to 1400 registered journalists and the organisation won a peer-community award for its coverage of the pandemic (Netzwerk Recherche, 2020).

In interviews conducted in January 2020, when the Covid outbreak had not yet been recognised as a pandemic, SMC Germany editors and management alike understood their role as that of a "service provider", helping journalists in need by providing "raw material" out of public view. However, in October of that year, in the midst of the pandemic, SMC Germany launched a media campaign called "Together for Fact News" directed at scientists, journalists and, to a lesser extent, the wider public. Starting from the premise that both independent scientific research and independent journalism are crucial for a liberal democracy, the campaign aimed to promote evidence-based science journalism by encouraging researchers to engage in public debate. SMC Germany founder Volker Stollorz explains: "In times when powerful actors spread lies, disinformation and confusion in their own interests, we are looking for researchers and journalists as ambassadors who stand up for reliable knowledge in public debates. And stand by it" (SMC Germany, n.d.). In a play on the confession of faith expressed in a religious creed, the campaign website features videos of well-known science journalists and scientists committing to a "knowledge creed" ("Wissensbekenntnis") (Together for Fact News, n.d.).

This transformation from service provider in the background, to campaign leader in the foreground, reveals the blurred boundaries entailed in categorising science communication as either journalism for the common good, or PR in the interests of actors in science. Susanna Priest (2018) makes the distinction between strategic and democratic science communication, the former serving the interests of those whom the communicator represents; the latter enabling citizens to reach informed decisions. SMC Germany's communication objectives are both strategic and democratic in nature. They are democratic in the sense that SMC Germany's activities are based on the premise that scientists have particular responsibilities in democracies; and they are strategic in that SMC Germany uses intentional forms of communication to attract attention to scientific topics and to shape opinions in favour of an ideal of objective reasoning that is in some respects ideological. As its campaign website puts it: "we love enlightenment" (Together for Fact News, n.d.) – an institutional commitment to a social programme, but one aimed at uplifting the democratic function of journalism in the public interest.

An important complicating factor is that SMC Germany's daily work strongly depends on its surrounding network and its management of trust-based relationships with actors in other organisations, each of which are pursuing their own interests (Broer and Pröschel, 2022). Any SMC needs to be considered credible and relevant by journalists, without whom there would be no media impact, but also requires the cooperation of press officers to gain access to new research and of individual scientists to make statements that can be disseminated as expertise. This dependency means that SMC Germany risks perpetuating existing medialising tendencies. These include a focus on the largest scientific journals, the preselection of publications in embargoed releases and a reliance on scientists who are willing and able to provide public commentary but who may not always be the most qualified to do so. In this regard, SMC Germany faces many of the constraints that science journalists deal with in traditional newsrooms.

Conclusion

Concerns about science PR often centre on the pursuit of individual or organisational interests in an environment in which visibility and media attention have become interlinked with economic and academic success (Taubert and Weingart, 2017). Science communication using PR practices can thus degrade into "institutional propaganda" (Weingart and Guenther, 2016: p. 4) and "science by soundbite" (Henderson and Kitzinger, 2007: p. 69), where journalism serves to legitimise particular science organisations rather than science in general (Peters et al., 2009: p. 31). Responding to these concerns, SMC Germany has explicitly aligned itself with the interests of journalism and, by extension, with the public interest. As the deputy editor puts it: "This is the difference: we're not an agency or press office for science. We are a science newsroom that wants to share information with others, so that they can share it again in their work. It is a service from journalists to journalists."

In an organisational analysis that affirms SMC Germany's identification with journalism, Christopher Buschow and co-authors (2022) characterise it as an organisation tasked with "field repair". It aims to improve the quality of journalism by undertaking journalistic activities in an alternatively financed model, thereby substituting for the journalistic work that commercial newsrooms are no longer able to perform. At the same time, SMC Germany is distinctive, these authors suggest, in combining this field repair work with what they call "field advancement", exploring and trialling journalistic innovations and technological interventions that can aid journalism practice. However, notwithstanding this close identification with journalism, in the critical moment brought about by the pandemic, the service functions of field repair and field advancement were supplemented with a campaigning function that highlighted the organisation's capacity for strategic communications akin to PR.

Ideas about science are tightly intertwined with democratic imaginations; visions about the ways that society should function and move forward (Ezrahi, 2003; Jasanoff, 1992; Miller, 2015). Science communication plays an important normative role in public discourse. Science communication practices, including science journalism, contribute to the development of particular kinds of societies and construct narratives about the role of science and technology for our futures (Felt and Wynne, 2007; Jasanoff and Kim, 2015). The establishment of SMC Germany and other new intermediary organisations situated between journalism and science can thus be seen as a reaction to what sociologist Silvio Waisbord calls the "collapse of the technocratic liberal dream" and its "modern project of disciplining knowledge by promoting the scientific model as the only legitimate knowledge" (2018: p. 1869).

In this sense, SMC Germany and similar organisations are not new: efforts to streamline communication between science and journalism go back to the roots of both fields in that technocratic dream. More than a century ago, news magnate Edward W. Scripps set up the independent Science Service news agency to address his concerns about the state of science news. His co-founder, biologist William E. Ritter, later recalled that Scripps was worried by the "vast quantity of misinformation being constantly spread abroad by our newspapers." In Scripps' vision, the only way to protect democracy was to make it more intelligent: "But since to be intelligent is utterly impossible without having much of the knowledge, the method, and the spirit of science, the only way to make democracy safe is to make it more scientific" (Ritter, 1926: p. 201).

SMC Germany recapitulates the same logic articulated by Scripps. Such historical cyclicity is simultaneously both disconcerting and reassuring. For the past century at least, the attribution to both journalism and science of the power to produce objective and truthful knowledge about the world has gone hand in hand with a deep-seated doubt about the ability of others to properly make sense of that knowledge – potentially to the detriment of society as a whole. As long as this is the basis of the science-media-society relationship, intermediary organisations will continue to be called on to walk a line between science PR and science journalism.

Note

1 If not otherwise referenced, quotes come from these interviews.

References

Baerns, B., 1985. *Öffentlichkeitsarbeit oder Journalismus? Zum Einfluss im Mediensystem*. Köln: Verlag Wissenschaft und Politik.

Bauer, M.W., Howard, S., Romo Ramos, Y.J., Massarani, L. and Amorim, L. (eds.), 2013. *Global Science Journalism Report: Working Conditions and Practices, Professional Ethos and Future Expectations*. London: SciDev.Net.

Blöbaum, B., 2017. Wissenschaftsjournalismus. In: Bonfadelli, H., Fähnrich, B., Lüthje, C., Milde, J., Rhomberg, M. and Schäfer, M. (eds.), *Forschungsfeld Wissenschaftskommunikation*. Wiesbaden: Springer VS, pp. 221–238.

Bornmann, L. and Mutz, R., 2015. Growth rates of modern science: A bibliometric analysis based on the number of publications and cited references. *Journal of the Association for Information Science and Technology*, 66(11), pp. 2215–2222.

Broer, I., 2020. Rapid reaction: Ethnographic insights into the Science Media Center and its response to the COVID-19 outbreak. *Journal of Science Communication*, 19(05), p. A08.

Broer, I. and Pröschel, L., 2021. *Das Science Media Center Germany: Ethnographische Einblicke in die Arbeitsweisen und Rollen eines Intermediärs zwischen Wissenschaft und Journalismus.* Hamburg: Verlag Hans-Bredow-Institut. https://doi.org/10.21241/ssoar.73542.

Broer, I. and Pröschel, L., 2022. Knowledge broker, trust broker, value broker: The roles of the Science Media Center during the COVID-19 pandemic. *Studies in Communication Sciences*, 22(1), pp. 101–118.

Brossard, D. and Scheufele, D.A., 2013. Science, new media, and the public. *Science*, 339(6115), pp. 40–41.

Buschow, C., Suhr, M. and Serger, H., 2022. Media work as field advancement: The case of Science Media Center Germany. *Media and Communication*, 10(1), pp. 99–109.

Christoph, C., 2009. *Textsorte Pressemitteilung: Zwischen Wirtschaft und Journalismus.* Konstanz: UVK.

Dunwoody, S., 2021. Science journalism: Prospects in the digital age. In: Bucchi, M. and Trench, B. (eds.), *Routledge Handbook of Public Communication of Science and Technology* (3rd ed.). Abingdon: Routledge, pp. 14–32.

Ezrahi, Y., 2003. Science and the postmodern shift in contemporary democracies. In: Joerges, B. and Nowotny, H. (eds.), *Social Studies of Science and Technology: Looking Back, Ahead.* Dordrecht: Springer, pp. 63–75.

Fahy, D. and Nisbet, M.C., 2011. The science journalist online: Shifting roles and emerging practices. *Journalism*, 12(7), pp. 778–793.

Felt, U. and Wynne, B., 2007. *Taking European Knowledge Society Seriously. Report of the Expert Group on Science and Governance to the Science, Economy and Society Directorate, Directorate-General for Research, European Commission.* Luxembourg: Office for Official Publications of the European Communities.

Foucart, S. and Horel, S., 2020. L'information scientifique sous tutelle d'une agence de communication? *Le Monde*, Sep., https://www.lemonde.fr/sciences/article/2020/09/22/l-information-scientifique-sous-tutelle-d-une-agence-de-communication_6053092_1650684.html.

Fox, F., 2012. Practitioner's perspective: The role and function of the Science Media Centre. In: Rödder, S., Franzen, M. and Weingart, P. (eds.), *The Sciences' Media Connection – Public Communication and Its Repercussions.* Dordrecht: Springer, pp. 257–270.

Franzen, M., 2012. Making science news: The press relations of scientific journals and implications for scholarly communication. In: Rödder, S., Franzen, M. and Weingart, P. (eds.), *The Sciences' Media Connection – Public Communication and Its Repercussions.* Dordrecht: Springer, pp. 333–352.

Franzen, M., 2014. Medialisierungstendenzen im wissenschaftlichen Publikationssystem. In: Weingart, P. and Schulz, P. (eds.), *Wissen – Nachricht - Sensation: Zur Kommunikation zwischen Wissenschaft, Öffentlichkeit und Medien.* Weilerswist: Velbrück, pp. 19–45.

Friedrichsmeier, A., Laukötter, E. and Marcinkowski, F., 2015. Hochschul-PR als Restgröße. Wie Hochschulen in die Medien kommen und was ihre Pressestellen dazu beitragen. In: Schäfer, M.S., Kristiansen, S. and Bonfadelli, H. (eds.), *Wissenschaftskommunikation im Wandel*. Köln: Halem, pp. 128–151.

Granado, A., 2011. Slaves to journals, serfs to the web: The use of the internet in newsgathering among European science journalists. *Journalism*, 12(7), pp. 794–813.

Gu, X. and Blackmore, K.L., 2016. Recent trends in academic journal growth. *Scientometrics*, 108(2), pp. 693–716.

Guenther, L. and Joubert, M., 2021. Novel interfaces in science communication: Comparing journalistic and social media uptake of articles published by the conversation Africa. *Public Understanding of Science*, 30(8), pp. 1041–1057.

Haran, J., 2012. Campaigns and coalitions: Governance by media. In: Rödder, S., Franzen, M. and Weingart, P. (eds.), *The Sciences' Media Connection – Public Communication and Its Repercussions*. Dordrecht: Springer, pp. 241–256.

Henderson, L., and Kitzinger, J., 2007. Orchestrating a science 'event': The case of the human genome project. *New Genetics and Society*, 26(1), pp. 65–83.

Hettwer, H., Schneider, M., and Zotta, F., 2013. *Explorationsphase zur Gründung eines Science Media Center (SMC) in Deutschland*. Dortmund: WPK.

Hoffjann, O., 2007. *Journalismus und Public Relations*. Wiesbaden: VS Verlag für Sozialwissenschaften. https://doi.org/10.1007/978-3-531-90744-4.

House of Lords, 2000. *Third Report: Science and Society*. Select Committee on Science and Technology, HL38, 1999-2000. London: The Stationery Office.

Jasanoff, S., 1992. The descent of Icarus: Science and the transformation of contemporary democracy, by Yaron Ezrahi. *American Political Science Review*, 86(1), pp. 233–234.

Jasanoff, S. and Kim, S.-H. (eds.), 2015. *Dreamscapes of Modernity: Sociotechnical Imaginaries and the Fabrication of Power*. Chicago: University of Chicago Press.

Kiernan, V., 2006. *Embargoed Science* (Illustrated edition). Urbana: University of Illinois Press.

Lemke, S., Brede, M., Rotgeri, S. and Peters, I., 2022. Research articles promoted in embargo e-mails receive higher citations and altmetrics. *Scientometrics*, 127(1), pp. 75–97.

Massarani, L., Neves, L.F.F., Entradas, M. and Bauer, M., 2021. *Global Science Journalism Report 2021: Working Conditions and Practices, Professional Ethos and Future Expectations*. London: SciDev.Net.

Matthews, J., 2014. Journalism. In: Maguire, J. S. and Matthews, J. (eds.). *The Cultural Intermediaries Reader*. London: Sage.

Miller, C.A., 2015. The epistemics of self-governance. In: Hilgartner, S., Miller, C. and Hagendijk, R., *Science and Democracy*. New York: Routledge, pp. 198–219.

Netzwerk Recherche, 2020. Sonderpreis 2020 für das Science Media Center (SMC). https://netzwerkrecherche.org/stipendien-preise/leuchtturm/sonderpreis-2020-fuer-das-science-media-center-smc/.

Peters, H.P., 2012. Scientific sources and the mass media: Forms and consequences of medialization. In: Rödder, S., Franzen, M. and Weingart, P. (eds.), *The Sciences' Media Connection – Public Communication and Its Repercussions*. Dordrecht: Springer, pp. 217–239.

Peters, H.P., et al., 2009. Medialisierung der Wissenschaft und ihre Relevanz für das Verhältnis zur Politik. In: Peters, H.P. (ed.), *Medienorientierung biomedizinischer*

Forscher im internationalen Vergleich: Die Schnittstelle von Wissenschaft & Journalismus und ihre politische Relevanz. Jülich: Forschungszentrum Jülich, pp. 9–43.

Phillips, D.P., Kanter, E.J., Bednarczyk, B. and Tastad, P.L., 1991. Importance of the lay press in the transmission of medical knowledge to the scientific community. *New England Journal of Medicine*, 325(16), pp. 1180–1183.

Priem, J., Piwowar, H.A. and Hemminger, B.M., 2012. Altmetrics in the wild: Using social media to explore scholarly impact. *ArXiv*, https://doi.org/10.48550/ARXIV.1203.4745.

Priest, S., 2018. Communicating climate change and other evidence-based controversies: Challenges to ethics in practice. In: Priest, S., Goodwin, J. and Dahlstrom, M.F. (eds.), *Ethics and Practice in Science Communication*. Chicago: University of Chicago Press, pp. 55–73.

Quandt, T. and Wahl-Jorgensen, K., 2021. The coronavirus pandemic as a critical moment for digital journalism. *Digital Journalism*, 9(9), pp. 1199–1207.

Raupp, J., 2017. Strategische Wissenschaftskommunikation. In: Bonfadelli, H., Fähnrich, B., Lüthje, C., Milde, J., Rhomberg, M. and Schäfer, M.S. (eds.), *Forschungsfeld Wissenschaftskommunikation*. Wiesbaden: Springer VS, pp. 143–163.

Rhein, L. and Lautenschläger, S., 2022. Wissenschaftskommunikation im Spannungsfeld von Wissenschaft, Politik und Medien: Zur Aushandlung von Gesprächsnormen in Pressekonferenzen und Polit-Talkshows. *Fachsprache*, 44(1-2), pp. 20–39.

Ritter, W.E., 1926. Science service and E. W. Scripps' philosophy of life. *The Science News-Letter*, 10(298), pp. 201–202. https://doi.org/10.2307/3902922.

Rödder, S., 2015. Science Media Centres and public policy. *Science and Public Policy*, 42(3), pp. 387–400.

Rödder, S., 2020. Organisation matters: Towards an organisational sociology of science communication. *Journal of Communication Management*, 24(3), pp. 169–188.

Schäfer, M.S., 2017. How changing media structures are affecting science news coverage. In: Jamieson, K. H., Kahan, D. M. and Scheufele, D. A. (eds.), *The Oxford Handbook of the Science of Science Communication*, Vol. 1. Oxford: Oxford University Press, pp. 50–59.

Scimex, 2014. SMC Global Network. *Scimex*, https://www.scimex.org/about/smc-global-network.

SMC, 2002. *Consultation Report*. Science Media Centre, Mar., https://www.sciencemediacentre.org/wp-content/uploads/2012/09/Science-Media-Centre-consultation-report.pdf.

SMC, 2012. Views from the Frontline. Science Media Centre, https://www.sciencemediacentre.org/wp-content/uploads/2012/10/SMC-Views-from-the-front-line-essays-on-the-SMC.pdf.

SMC Germany, n.d. Together for Fact News. Science Media Center Germany, https://www.sciencemediacenter.de/das-smc/together-for-fact-news/.

SMC Germany, 2022. Wie Wir Uns Finanzieren. Science Media Center Germany, https://www.sciencemediacenter.de/das-smc/wie-wir-uns-finanzieren/.

SMC Global, 2012. *Charter of Guiding Principles for Science Media Centres*. SMC Global, https://www.smcglobal.org/resources.

St Louis, C., 2013. Don't let the PR industry annex science journalism. *Research Fortnight*, 16 Jan., p. 22.

Suhr, M., Buschow, C. and Serger, H., 2022. Organisationsinnovationen im Journalismus: Das Science Media Center Germany als Prototyp einer neuartigen

Unterstützungsinfrastruktur für den Journalismus. In: Pranz, S., Heidbrink, H., Stadel, F. and Wagner, R. (eds.), *Journalismus und Unternehmenskommunikation*. Wiesbaden: Springer, pp. 147–162.

Taubert, N.C. and Weingart, P., 2017. Changes in scientific publishing: A heuristic for analysis. In: Taubert, N.C. and Weingart, P. (eds.), *The Future of Scholarly Publishing: Open Access and the Economics of Digitisation*. Cape Town: African Minds, pp. 1–33.

Together for Fact News, n.d. Home. Science Media Centre Germany, https://www.together-for-fact-news.de/.

Väliverronen, E., 2021. Mediatisation of science and the rise of promotional culture. In: Bucchi, M. and Trench, B. (eds.), *Routledge Handbook of Public Communication of Science and Technology* (3rd ed.). Abingdon: Routledge, pp. 129–146.

Waisbord, S., 2018. Truth is what happens to news. *Journalism Studies*, 19(13), pp. 1866–1878.

Wakeford, T. 2001. New Stalinism in the Labour lab. *The Guardian*, 15 Jun., https://www.theguardian.com/education/2001/jun/15/highereducation.comment.

Ware, M. and Mabe, M., 2015. *The STM Report: An Overview of Scientific and Scholarly Journal Publishing*. Oxford: International Association of Scientific, Technical and Medical Publishers.

Weingart, P., 2001. *Die Stunde der Wahrheit? Zum Verhältnis der Wissenschaft zu Politik, Wirtschaft und Medien in der Wissensgesellschaft*. Weilerswist: Velbrück Wissenschaft.

Weingart, P. and Guenther, L., 2016. Science communication and the issue of trust. *Journal of Science Communication*, 15(5), C01. https://doi.org/10.22323/2.15050301.

Weingart, P., Joubert, M. and Connoway, K., 2021. Public engagement with science: Origins, motives and impact in academic literature and science policy. *PLOS ONE*, 16(7), p. e0254201. https://doi.org/10.1371/journal.pone.0254201.

Williams, A. and Gajevic, S., 2013. Selling science? *Journalism Studies*, 14(4), pp. 507–522.

6

THE OVER-RELIANCE ON FOREIGN SCIENCE NEWS IN DEVELOPING COUNTRIES

Causes, consequences and solutions

An Nguyen and Minh Tran

For many, if not most, farmers growing grapefruits in Vietnam, the BBC and the *Daily Mail* must have always seemed like abstract entities from a far-flung place, far removed from their everyday lives. That was, until the summer of 2007. As the farmers' newly harvested grapefruits landed in local markets around the country, news broke in several Vietnamese newspapers that a new American study had found that grapefruit consumption could cause breast cancer among post-menopausal women. These stories were translated and adapted from two British stories – "Grapefruit link to cancer" on the BBC News website and "Eating grapefruit can increase breast cancer risk by a third" in the *Daily Mail.* Soon afterwards, the local grapefruit market collapsed, with consumers boycotting the fruit and its price plummeting to 10%, or even less, of what it had been before the news broke.

Leaving aside errors in the UK reports that were not heeded by those who regurgitated them in Vietnamese (which we will discuss later), a quick fact check would have revealed that the grapefruit sample in the American study was a species that was not grown in Vietnam. When this was later discovered, four Vietnamese news titles were fined for carelessly publishing the stories. By then, however, the damage had been done, especially to the many poor local farmers who had suffered huge economic losses from the media-induced boycott. According to one estimate, one single province in the Mekong Delta lost VND 100 billion to those stories (around USD 600,000 at the time) – a hefty price tag for Vietnamese farmers.

This saga, uncovered in our research into Vietnamese science journalism over the past decade, is one of many examples showing how 24/7 global news can cause immediate damage to seemingly disconnected communities around the world. What is rather ironic here is the nature of the news subject –

DOI: 10.4324/9781003326724-9

a scientific study that aimed to contribute to the welfare of humanity. Of course, the research itself was not to blame: the culprit lay in the way it was represented in the global news system, first in English-language news outlets and then, through translation, in Vietnamese ones. Of interest in this chapter is the widely observed, but largely uncharted, issue at the heart of global science communication: the heavy dependence of the media of the Global South on science news and information sources from the Global North. From Africa and the Middle East to South America and developing parts of Asia, a large proportion of science news consumed in the Global South is translated or, at best, synthesised from foreign sources, especially media outlets based in the North. Although this science news structure bears many long- and short-term implications for local and global development, it has rarely been studied in any depth.

In what follows, we take Vietnam as a case study through which to examine the nature of the over-reliance on foreign science sources by the news media of the Global South. We do so by presenting data from a content analysis of science news in Vietnamese online media during 2017 and 2018, and by drawing on in-depth interviews with over 40 Vietnamese science journalists and news executives, conducted between 2012 and 2020.[1] We begin by first examining the current state and causes of the over-reliance on foreign news, before considering its short-term and long-term consequences for countries in the Global South and the prospects of more independent science journalism emerging in the future. The reliance on foreign news is a double-edged sword: while it helps enhance general awareness and understanding of global science developments among people in the Global South, it can have negative consequences that extend far beyond short-term damage, like the grapefruit story. The most significant of these are the colonial legacies embedded and reinforced in the science news dependency, which can be seen as part of a neo-colonial structure of knowledge dependency. The situation, we conclude, can be improved, at least partially, with a reform of journalistic approaches to science topics and a strong capacity-building agenda for science journalism in the Global South.

The absence of 'homemade' science news in the Global South

A substantial body of academic and policy research has shown that, even in the easy publishing and consuming environment of digital platforms, global news flows predominantly in one direction only, from the Global North to the rest of the world (Bromley and Slavtcheva-Petkova 2019; Dimitrova, 2021; Hachten and Scotton, 2016; Segev, 2016; Williams, 2011). With their economic and geopolitical importance and their well-resourced media systems, the world's major powers are also its major news powers. News and information flow almost vertically from the centre (developed countries in the Global North) to the periphery (developing ones in the Global South). Over time, this creates a structure of dependency that bears widespread socio-cultural, economic and political ramifications for poorer societies.

A similar one-way information infrastructure, although under-researched, has been seen in the realm of global science communication. The limited literature points to an excessive reliance on foreign science information sources, especially Northern news media, across the Global South (Nguyen and Tran, 2019). In Vietnam, for example, one of the most recurrrent issues that came up in our interviews with science journalists is the lack of home-made science news. Participants lamented that most international science stories are entirely duplicated or extracted from high-ranked news outlets of the North, with little link to local development. This is clearly reflected in our content analysis of Vietnamese online news coverage of genetically modified organisms (GMOs) during 2017 and 2018. As seen in Table 6.1, more than half of our 401 sampled stories (56%) about GMOs concern events outside Vietnam, primarily research achievements in scientifically advanced countries (Tran, 2021). Of the foreign science stories, 93% contained no explicit link to local contexts. Most of these were merely translated or edited down from materials that were published by the media and, to some extent, science journals and institutions in the developed world.

The Vietnamese situation is typical of other developing countries, where news outlets and science insitutions from the developed world are the most frequently used science information resources (Comfort et al., 2020; Evans, 2015; Han et al., 2017; Tagbo, 2010; Takahashi and Meisner, 2012). In Africa, for example, most science news is translated from European or North American media, with little or no local context and with very rare references to local science communities (Joubert, 2007; Shanahan, 2006; Tagbo, 2010).

TABLE 6.1 Geographical origins of GMO news stories in the Vietnamese media (2017-2018).

		Frequency	*Percentage (%)*
Geographical scope of event/issue *(Based on entire sample, N = 401)*	International	224	55.9
	Domestic	147	36.7
	General	30	7.5
Link to local context *(Based on the number of stories with an international scope, N = 224)*	Yes	16	7.1
	No	208	92.9
Originality *(Based on the number of stories with an international scope, N = 224)*	Mainly original reporting	18	8.0
	Mainly translated reporting	206	92.0

Source: Tran (2021).

In Latin America, Luisa Massarani and colleagues, analysing newspapers from ten countries, find that they pay far more attention to the so-called 'First World' than to local and regional science events and issues (Massarani et al., 2005; Massarani and Buys, 2007). For instance, a notable 70% of science and technology stories published in two prestigious Brazilian newspapers, *O Globo* and *Folha de S. Paulo*, were related to things that happen in developed countries. In most cases, the researchers conclude, science news is "republished without enough concern in adequating to local reality" (Massarani et al., 2005: p. 5). In Chile, interviews with 70 local journalists show that those covering science use European and American news outlets as the main source of references (Lorena et al., 2014). From a slightly different perspective, Mercado (2012) observes that Argentine newspaper coverage of topical science issues, such as climate change, is mainly found in the foreign or international affairs sections, rather than the environment or science sections.

What are the causes?

Research into the reasons for such over-reliance of the Global South on the Global North's science news and information sources is at best scattered and offers little depth. The most often mentioned cause is that newsworthy science achievements rarely originate at local laboratories or universities in the South (Joubert, 2007). As one Vietnamese science correspondent commented:

> There are not many scientific achievements in Vietnam. Funded research topics are repeated in some fashion year after year. Of course, there must be something 'new' in the lab, but there are not many breakthroughs and practical applications for us to cover.

When the rare cases of research successes happen in the South, they are often published in international journals in the first instance. As Cameroon journalist Buma Gana has observed, "African scientists are unknown at home" (Shanahan, 2006: p. 392) since good African academic research is normally published in the North rather than within its country of origin. The long-term consequence of this is a general perception among journalists that local science developments would be less reliable and offer less news values than those from developed countries (Demeter, 2019; Dutt, 2012; Víctor, 2019). As one Vietnamese reporter observed: "It is unspoken but the [poor] quality of science research in Vietnam is widely recognised among journalists."

The repackaging of science news from the North is prevalent not only in new science discoveries but also in global science issues with huge implications for the South. One Vietnamese editor was frustrated when describing his team's struggle to find scientists to provide a localised assessment of the potential impact of controversial scientific advances, such as GMOs. His criticism of local scientists was scathing:

> I have met scientists with rather outdated knowledge in the areas where they claim to be experts, especially controversial science issues. Sometimes, I can anticipate what they are going to say even before the interview, as they keep giving the same answers every year. I even feel that some are not as updated as me in issues such as GMO. Who in Vietnam cares enough about GMO these days to do some original research in this area? No one. No institution. All that they know comes from foreign labs and publications.

Where there is local expertise, again, access to experts proves to be very difficult, in part because scientists are reluctant – and not required or trained – to bring their research to the public through the media (Elia, 2019; Lorena et al., 2014; Nguyen and Tran, 2019). In many places, bureaucratic interferences and political control are formidable impediments to scientists' willingness and readiness to appear in the media. But perhaps the most fundamental issue is that most scientists do not see the value of making their work known beyond academia or their voices heard in the wider community. In Vietnam, for example, one journalist reported that scientists "are not open to the media because they see their job as doing research rather than communicating science to the public and they have no incentive to do it." Another cited a case in which he tried hard, but ultimately failed, to obtain an interview with a Vietnamese scientist who was among Thompson Reuters' 100 most cited researchers during 2014-2017. The journalist interpreted this experience as a reflection of scientists' inability to connect to society and contextualise their own research: "Genuine scientists would rather work silently and tend to forget that science is only meaningful when it is juxtaposed with the social, economic and political issues of the time." Similar things happen in the Arab world; despite the promising status of science and technology in their national development visions and policies, for example, Arabian science reporters still face tremendous challenges in finding local experts who have adequate knowledge and enthusiasm to communicate with the press (Alhuntushi and Lugo-Ocando, 2022).

The long-term consequence of all this is a general perception among journalists that local science developments are less reliable and offer less news value than those from developed countries (Demeter, 2019; Dutt, 2012; Víctor, 2019). This becomes a vicious circle: scientists and scientific institutions do not engage with the media, further reinforcing journalists' perceptions that there is a lack of local experts, which over time leads the journalists to make little effort to trace and find local experts in their news gathering.

The lack of local achievements and experts aside, however, there are more fundamental problems with the news industry that contribute to the South's excessive reliance on Northern science news sources. For one thing, newsrooms of the South are not comfortable with – much less friendly towards – the supposedly "hard and dull" science topics. Despite the importance of science for developmental causes and processes, the media often treat it as a "nice to have" rather than a "must have" in the news agenda. Investment in domestic science topics is often skewed towards "news you can use", such as health and lifestyle stories and environmental issues. Added to this is the general shortage of specialist science journalists and, in many places, the political control of science reporting, especially when it involves sensitive science and development policies (Nguyen and Tran 2019). As such, instead of the active pursuit and rigorous examination of science sources, journalists often conceptualise and develop stories from available research reports, press releases, government announcements or PR events. In that context, news translation is surely a quick and easy fix to fill up news space and airtime with science stories in order to maintain a comprehensive news agenda and attract certain segments of the audience without upsetting the political authorities.

Short-term impacts on daily life

Translated science news plays a crucial role in the Global South. It equips local audiences, including lay publics and decision-makers, with knowledge and updates about new science developments around the world. A long-standing problem in the South is the general lack of effective non-news science communication channels, such as science festivals and museums (Nguyen and Tran, 2019). In that context, it would be difficult to envision how the peoples and nations of the South would be able to keep up with the rapid, ceaseless pace of contemporary scientific discoveries and their socio-economic and political implications without access to foreign-sourced science news.

However, when most science news in developing countries is based on translation work by those with foreign language skills rather than science reporting experience, there are many concerning consequences, both in the short term and the long term. Rather than "select material from various sources and then weave these snippets into an independent and unique storyline" (Haapanen and Perrin, 2019: p. 19), translators tend to duplicate foreign stories that are devoid of local peculiarities. Without a general understanding of scientific methods and the ways in which science impacts on the contemporary world, these journalists are unable to critically assess original science sources or interpret the meanings of scientific developments accurately and proportionately. Such news coverage occurs especially frequently in appealing topics with immediate implications for daily life, such as health risks and benefits. Without cross-checking and localisation, there is a high risk of misleading information that, in some cases, can cause severe damage to the daily life of local communities.

News about the Covid-19 vaccine is a case in point. In their examination of factors fostering vaccine hesitancy in Pakistan, for example, Shafiq Ahmad Kamboh and colleagues highlighted two incidents of costly, but inadequate, translation that exemplify this concern. First, in August 2021, many Pakistanis started to abandon the Moderna vaccine after the Urdu newspaper, the *Daily Express*, compressed and reduced a detailed US analysis of the vaccine's higher likelihood to cause myocarditis compared to the Pfizer vaccine, into a short straight news item without any explanatory information. Second, sensational reports about two Americans dying hours after getting Covid-19 vaccines lacked details on the causes of deaths. The researchers suggest that these may have engendered public fears of vaccination. Many Urdu-language newspapers did not even fact-check adverse vaccination cases that they had gathered from foreign sources (Kamboh et al., 2022).

Here, again, Vietnam offers some specific insights. The science journalists we interviewed expressed grave concerns over the influx of inaccurate, confusing and sensationalised news about science-related risks from foreign media. One editor lamented that the Vietnamese online media, thanks to their limitless space, would publish foreign research into health risks with little attention given to their ethical duties towards their audiences:

> There's little sense of journalistic responsibility here. Online media are inundated with international research on whether coffee, tea, beer and so on are safe or unsafe. They would publish anything that comes their way as long as they see an audience interest in it and as long as it is not against the law or the state's propaganda line. They would publish such stories simply because they wouldn't lead to any death.

The result, as another observed, is that "health news in Vietnam, instead of helping people live a better, more healthy life, is like a smokescreen that creates confusion, panics and chaos". In his words:

> Here, if you follow the news, you are likely to find that you might die of cancer some day because of the daily food you happen to eat. And the way for many to move on is just to ignore these scare stories and keep going with their eating habits. People are paralysed.

Vietnam has seen many occasions in which the local economy has been severely afflicted due to false risk claims in the news and the grapefruit research case at the start of this chapter is one such example. Another is the way the Vietnamese media responded to Chinese and Northern media reports about melamine-contaminated milk in 2008. Following incidents in which some melamine-adulterated formula milk caused serious ill health (including kidney stones) among Chinese children, there was a flood of melamine-related risk coverage in Vietnamese media. Confused consumers, unable to distinguish melamine-contaminated from non-contaminated milk, opted for the safe route: turning their backs on all local dairy products. It was not until a month or so after the first news broke in China that Vietnamese dairy products were certified to be melamine-free. By that time, however, dairy farmers and producers had already paid a hefty price. Reflecting on this, a science journalist in Ho Chi Minh City said:

> In a country where science literacy is low, ordinary people place trust on information from the news media. But those who translated and edited these stories had no knowledge at all about the scientific issues around melamine.

More recent news coverage seems to suggest that outlets have not learned the lessons of such incidents. In April 2019, for example, Japanese authorities in Osaka recalled over 18,000 bottles of Chinsu chilli sauce imported from Vietnam because they contained benzoic acid over the level allowed by Japanese authorities. Several newspapers did take care to point out that the amount of benzoic acid in Chinsu sauce was deemed high by Japanese criteria but was accepted, not only by the Vietnamese authorities, but also by international food safety regulators such as the US FDA and CODEX. These newspapers, however, were the exception rather than the norm. The possibility of benzoic acid causing cancer was exaggerated in most of the news, and the subsequent media frenzy around the threat of benzoic acid to human health was intense enough to frighten many consumers who were not well equipped to engage critically in the coverage. Needless to say, a local boycott of Chinsu sauce followed.

In fact, many common mistakes found in science journalism of the North have been repeated and worsened in the science news coverage of the South. Again, the grapefruit story offers a case in point. Here, the mistake was not just around the unrealised difference between the grapefruit species in the American study and those grown in Vietnam, even though this was the verdict delivered by Vietnamese authorities to the four news organisations that were fined. Missing in that verdict was the exaggeration, intentional or unintentional, of the research findings in the original stories by the BBC and the *Daily Mail*. First, they mistakenly treated an observational correlation as a causal relationship. The study found the risk of breast cancer was 30% higher among post-menopausal women who ate a quarter of a grapefruit or more per day than among those who did not. In other words, there was a correlation between breast cancer and grapefruit consumption among these women, but that did not mean the latter was the cause of the cancer as it seemed from the original UK stories. Second, these reports were made more alarming by describing an impressive *relative* risk (the percentage of risk increase) without mentioning the rather small *absolute* risk (the number of cases). Third, they reported on the inconclusive results of a single study without adequate warning that these were not time-tested established knowledge and further research was required before a reliable conclusion could be reached. None of the Vietnamese translators (nor, later, the media regulators) were able to spot, let alone avoid, these underlying errors.

The reinforcement of the colonial science order

Far more serious in the long term is the incompatible nature of much of the scientific knowledge that flows into developing countries. Science is a global public good but for that good to work to the benefit of humanity, it needs to be used and implemented in tandem with local needs, cultures, economics and politics. In our interviews with Vietnamese journalists, while it was agreed that foreign stories might help update and increase scientific knowledge and understanding among certain segments of the public as well as informing local policy makers, there was a real unease and frustration about the fact that they were irrelevant or even incomprehensible to most people. As the interviewees argued, science and technology cannot succeed in any social system if local and indigenous factors are not integral to the adoption process. One reporter cited as an example Vietnamese use of Japanese safety standards for nuclear electricity plants. "Scientists have long proven that Vietnam's geological conditions, unlike Japan's, are quite stable," he said. "So, is it suitable to evaluate Vietnamese nuclear applications based on Japanese standards?" That has not included, he added, the fact that foreign standards are not always right. In this case, the Japanese vastly underestimated the actual risks of their nuclear energy production facilities, which they did not realise until the Fukushima earthquake and tsunami in 2011.

Similar concerns can be found in other parts of the South (Dutt, 2009; Dutt and Garg, 2012; Verghese, 2016). The regurgitation of science news from the North is especially problematic when it comes to development policies related to the major global science issues of our time – such as climate change, genetic engineering, artificial intelligence, nanotechnology, stem-cell research, green energy, neuroscience and so on. Debates in these areas of science are largely initiated in the North and exported to the South via various routes – such as global governance bodies (e.g., the UN and its agencies, World Bank, IMF and WHO), multi- and trans-national corporations (through their financial and technical transfer mechanisms), foreign aid agencies, international NGOs and charities, research funding councils and cultural exchange bodies. But in all cases, as already discussed, the media hold the most far-reaching influence in informing and enlightening local publics and decision-makers of the South (Nguyen and Tran, 2019). Without specialist skills and knowledge in translating and reporting such complex science, however, news outlets can easily duplicate and align with Northern worldviews, standards, policies and debates. The situation, of course, is worsened by the above challenges for journalists of the South to find relevant local voices (Han et al., 2017; Joubert, 2007; Tolan, 2007) and also by the sidelining of science debates in the public agenda in many countries (Nguyen and Tran, 2019).

We all know, for example, that climate change cannot be addressed without a full consideration of its disproportionate consequences on the peoples of the South and of the developmental needs of the South. Yet the Southern agenda for climate change is frequently framed from Northern viewpoints and, as such, tends to mirror Northern approaches (Hase et al., 2021; Keller et al., 2020). One Vietnamese science reporter cited green energy as an example of the divergence between what the North wants to pursue and what the South needs:

> For a rapidly developing country like Vietnam, energy is always a contentious pressing issue. Hydroelectricity capacity has been used up. Nuclear power is opposed. Solar power is green but takes time and money to develop. If Vietnam followed everything proposed by international environment groups, it would have to stop exploiting natural resources such as coal. But would we be able go back to using kerosene lamps in place of electric light in the next five or ten years to wait for solar power? There needs to be a balance between development needs and environmental considerations as well as a locally compatible roadmap towards green energy.

Here, a double-layered dependency emerges that serves to reinforce the colonial order of science and knowledge. On one level is a news sphere in which the "information haves" of the North, especially the United States and the United Kingdom, dictate global terms and conditions to the "information have-nots" of the South. On another level is the centuries-long dominance of the North's socio-economic and political needs, epistemic reasoning and theoretical paradigms in scientific research. Both are implicated in the colonial system in which the rationality-driven, truth-seeking practices of science and journalism were associated with the image of the North as "civilised societies". The logical outgrowth of this was that the North sought to expand its socio-technological and scientific expertise to help "barbaric societies" become civilised, or at least less uncivilised. Much of that expansion was through the international news system, which was itself created to serve colonialism (Williams, 2011). Over time, accepted knowledge from the North – especially knowledge through science and journalism – became universalised, transforming itself into power and its producers into authorities.

As historian Rohan Deb Roy (2018) has argued, as power became inseparable from knowledge, it reinforced the old order dominated by the 'superior'

North and extended it well into contemporary science. In particular, as both the practice of science and a science-literate public are culturally attached to only the North, those outside the North were considered underdeveloped and 'intellectually inferior': they needed to progress through the gracious gift of science that European empires (and later the United States) brought to them. In other words, they "deserved and needed to be colonized". As Roy puts it:

> In these ways and more, Europe's leaps in science and technology during this period both drove and were driven by its political and economic domination of the rest of the world. Modern science was effectively built on a system that exploited millions of people. At the same time it helped justify and sustain that exploitation, in ways that hugely influenced how Europeans saw other races and countries.
>
> (Roy, 2018)

As European empires declined in the post-World War II era, the civilisation discourse associated with the concept of colonialism was replaced by the development paradigm. This was invented as a tool for the establishment of a new geopolitical order by the new superpower, the United States. This paradigm, again, was based on a presumption of Northern superiority. Rationality, scientific advancement and technological innovation – along with values and practices such as the free-market economy and liberal journalism – continue to be the justification for the 'developed' First World to determine and control the progress and advancement of the 'under-developed' Third World (Lugo-Ocando and Nguyen, 2017). Thus, although empires may have disappeared, Roy argues that "colonial legacies continue to shape trends in science today". This is because the cultural biases and inequalities that these legacies imposed, as well as the old scientific hierarchy, are here to stay – in the form of annual university rankings by Western standards, the dominance of United States and Western Europe academic journals and so on. As "excellence in science and technology are a euphemism for significant funding, infrastructure and economic development", a neo-colonialism order comes into play, claims Roy. Countries of the Global South "are seen either as playing catch-up with the developed world or as dependent on its scientific expertise and financial aid." Seen in this context, the global news system, with its continued reliance on a one-way flow from North to South, plays an instrumental role in reinforcing and perpetuating the neo-colonialist scientific order.

Conclusion: Where to from here?

Notwithstanding the challenges outlined above, science reporting in the Global South has improved over time and changes have been seen in some emerging economies. Many journalists of the South now seem to be acutely conscious of the limitations of the North-dominated science discourse for their national development processes. This is evident in the responses of our Vietnamese interviewees, such as the above comment about the international green energy agenda. It is even more evident in emerging economies – including China, India, South Africa, Brazil and some parts of the Arab world – whose developmental strategies have resulted in remarkable science and technology successes. In China, for example, there has been an increasing preference for domestic rather than foreign science sources (Soete, 2016). In India, longitudinal studies of the science and technology coverage in English-language dailies show that local sources have gained an increasingly stronger profile alongside foreign voices (Dutt and Garg, 2000; 2012). Similarly, Brazil has seen more local science developments in print and television news (Ramalho et al., 2012).

Despite such positive changes, however, these countries still see foreign voices trumping local peculiarities and priorities in the news framing of science and development policies (Alhuntushi and Lugo-Ocando, 2022; Hase et al., 2021; Keller et al., 2020). It remains challenging for journalists in these countries, and even more so for those from less developing parts of the Global South, to change the status quo. Due to the many internal and external impediments to science journalism in the South, as well as the historical dependency structures discussed above, it is unlikely that the dominant one-way flow of science news and information from North to South will disappear in the foreseeable future. Its instrumental role in promoting and reinforcing Northern-led worldviews in global science development and policies is likely to continue, not always with negative consequences but usually with a loss of scientific – and with that socio-economic – sovereignty on the part of the South.

Indeed, in one respect, the positive changes in emerging economies noted above speak to a major issue that this chapter has identified: the link between the (un)availability of local scientific achievements and voices on the one hand, and the media's (un)interest in sourcing news from local scientists and their institutions on the other. With these countries moving into new, cutting-edge science and technology and emerging as new science powerhouses, their local experts have been featured more frequently in the news. For example, Brazilian science, which enjoys an investment rate per gross domestic product (GDP) that is relatively equal to that of Canada or the UK, has reached an advanced stage where it has much to offer the local media (UNESCO, 2010; Soete, 2016). In fact, Brazil was the world's 13th largest publisher of science papers on the Thomson Reuters Science Citation Index between 2011 and 2016 and the 15th highest publisher on Scimago's journal and country rankings for the period between 1996 and 2017. From this vantage point, the current over-reliance on Northern science sources of Southern media could be expected to decrease as a nation's own scientific research matures and advances. The less science a country has to catch up with and the more sovereignty it has in research, the more scientific voices will appear in the media, both foreign and domestic.

That said, there is much that the media of the South could do themselves to improve the current imbalances in the flow of global science news and information. External impediments to journalists' professional practice – such as poor and difficult access to local expertise or the political control of science and development debates – are somewhat beyond their control, but internal barriers are not insurmountable if news outlets are in the business of supporting the long-term development of their countries. At the top of any potential reform agenda should be an overhaul of traditional journalistic approaches to science and technology, with a strategic vision to develop a science-friendly news culture. Such a culture would need to treat science as an essential news beat, seeing it as relevant, as meaningful and as impactful as economics or politics. It would find news values not only in "breakthrough" discoveries, scary risks and marvellous benefits but also, and more importantly, in the connection between scientific research and daily life and in the developmental aims of developing societies. Ideally, that should include localising the global (linking science findings, standards, strategies and policies from other countries to local considerations) and globalising the local (linking local science to relevant research interests, discoveries and issues elsewhere).

This might seem like wishful thinking given the current decline of journalism's economic health, which has led newsrooms, including many in the North, to abandon science news teams altogether. The good news, as we heard from interviews with news editors and executives in six Southeast-Asian countries, is that they have a genuine willingness to invest in science topics (Nguyen, 2014). None of the editors we interviewed denied that science has not been given due importance in their news output. They admitted that there was a perception among editors that science lacks newsworthiness. However, they categorically refuted the perception that they do not care about science, citing various efforts to increase the amount of science coverage in their news output. The editor-in-chief of a Vietnamese daily, for example, recalled that he had once run a science section, only to have to discard it when it struggled to thrive. "The section was more a patchy content space than something solid and attractive," he said. "We were stuck because we had no relevant expertise to improve it." The following comment from another editor summarises the situation:

> I think if reporters know how to report science in an interesting way, science topics could appear in every page, including the front and second page. But because we don't know how to deal with science, we have missed many opportunities to make it popular, interesting and attractive to audiences.

Research elsewhere has also pointed to the lack of specialist expertise as a crucial contributor to the dominance of foreign science in the Global South's media (e.g., Lorena et al., 2014). In short, the bottleneck is not a 'don't care' attitude, but rather a 'don't know' issue.

It follows from this that any effort to alleviate the negative effects of the South's overdependence on Northern science sources should start from a strong agenda for professional capacity building for journalists in the South to cover science. That includes, of course, the continued enhancement of science reporting expertise for specialist beat journalists that professional associations such as the World Federation of Science Journalists have focused on for years (see Chapter 3). But it is also time to equip those on the foreign news desks with the fundamental knowledge and skills in sourcing, evaluating and re-interpreting foreign research materials, especially those from the global news system. No less importantly, editors and reporters need to learn to replace the traditional disconnection between foreign news and science journalists with a close collaboration and coordination through newsroom-wide workflows. Such capacity building will not solve every problem, but at the very least it will create the conditions for many other positive changes to come over time.

Note

1 Methodological details can be found in our recent work: Nguyen (2014), Tran (2021) and Tran and Nguyen (2023). Where not otherwise referenced, quotes come from our interviews. This chapter reuses some quotes published in Tran and Nguyen (2023).

References

Alhuntushi, A. and Lugo-Ocando, J., 2022. Articulating statistics in science news in Arab newspapers: The cases of Egypt, Kuwait and Saudi Arabia. *Journalism Practice*, 16(4), pp. 702–718.

Bromley, M. and Slavtcheva-Petkova, V., 2019. *Global Journalism: An Introduction.* London: Red Globe Press.

Comfort, S.E., Tandoc, E. and Gruszczynski, M., 2020. Who is heard in climate change journalism? Sourcing patterns in climate change news in China, India, Singapore, and Thailand. *Climatic Change*, 158, pp. 327–343.

Demeter, M., 2019. The winner takes it all: International inequality in communication and media studies today. *Journalism and Mass Communication Quarterly*, 96(1), pp. 37–59.

Dimitrova, D. (ed.), 2021. *Global Journalism: Understanding World Media Systems.* London: Rowan and Littlefield.

Dutt, B., 2009. An overview of science and technology coverage in Indian English-language dailies. *Public Understanding of Science*, 9(2), pp. 123–140.

Dutt, B., 2012. Biomedical research coverage in English-language Indian newspapers. *Journal of Scientometric Research*, 1(1), pp. 71–78.

Dutt, B. and Garg, K.C., 2012. S&T coverage in English-language Indian dailies. *Journal of Science Communication*, 11(3), p. A01. https://doi.org/10.22323/2.11030201.

Elia, E.F., 2019. Media coverage of climate change information in Tanzania. *Global Knowledge, Memory and Communication*, 68(4/5), pp. 258–274.

Evans, S., 2015. Journalistic norms, cultural values, and coverage of climate change in the Philippines. *Environmental Communication*, 10(4), pp. 492–507.

Haapanen, L. and Perrin, D., 2019. Translingual quoting in journalism: Behind the scenes of Swiss television newsrooms. In: Davier, L. and Conway, K. (eds.), *Journalism and Translation in the Era of Convergence*. Amsterdam: John Benjamins, pp. 15–42.

Hachten, W. and Scotton, J., 2016. *The World News Prism: Challenges of Digital Communication* (9th ed.). London: Wiley.

Han, J., Sun, S. and Lu, Y., 2017. Framing climate change: A content analysis of Chinese mainstream newspapers from 2005 to 2015. *International Journal of Communication*, 11, pp. 2889–2911.

Hase, V., Mahl, D., Schäfer, M.S. and Keller, T.R., 2021. Climate change in news media across the globe: An automated analysis of issue attention and themes in climate change coverage in 10 countries (2006–2018). *Global Environmental Change*, 70(September), https://doi.org/10.1016/j.gloenvcha.2021.102353

Joubert, M., 2007. South Africa: Building capacity. In: Bauer, M.W. and Bucchi, M. (eds.), *Journalism, Science and Society*. London: Routlege, pp. 251–253.

Kamboh, A.S., Ittefaq, M. and Sahi, A.A., 2022. Journalistic routines as factors promoting COVID-19 vaccine hesitancy in Pakistan. *Third World Quarterly*, 43(1), pp. 278–287.

Keller, T.R., Hase, V., Thaker, J., Mahl, D. and Schäfer, M.S., 2020. News media coverage of climate change in India 1997–2016: Using automated content analysis to assess themes and topics. *Environmental Communication*, 14(2), pp. 219–235.

Lorena, B.V., Evelyn, N. and Roberts, R., 2014. Communicating science in Chile. Problems in journalism training and scientific communication. *13th International Public Communication of Science and Technology Conference*, (November). Salvador, Brazil.

Lugo-Ocando, J. and Nguyen, A., 2017. *Developing News*. London: Routledge.

Massarani, L. and Buys, B., 2007. Science in the press in nine Latin American countries. *Brazilian Journalism Research*, 3(2), pp. 77–96.

Massarani, L., Buys, B., Amorim, L.H. and Veneu, F., 2005. Science journalism in Latin America: A case study of seven newspapers in the region. *Journal of Science Communication*, 4(3), p. A02. https://doi.org/10.22323/2.04030202

Mercado, M.T., 2012. Media representations of climate change in the Argentinean press. *Journalism Studies*, 13(2), pp. 193–209.

Nguyen, A., 2014. *The Current Status of Science Journalism in Southeast Asia*. Unpublished report to the World Federation of Science Journalists.

Nguyen, A. and Tran, M., 2019. Science journalism for development in the Global South: A systematic literature review of issues and challenges. *Public Understanding of Science*, 28(8), pp. 973–990.

Ramalho, M., Massarani, L. and Polino, C., 2012. From the laboratory to prime time: Science coverage in the main Brazilian TV newscast. *Journal of Science Communication*, 11(2), p. A02. https://doi.org/10.22323/2.11020202.

Roy, R.P., 2018. Decolonise science – time to end another imperial era. *The Conversation*, 5 Apr., https://theconversation.com/decolonise-science-time-to-end-another-imperial-era-89189.

Segev, E., 2016. *International News Flow Online: Global Views with Local Perspectives*. New York: Peter Lang.

Shanahan, M., 2006. Fighting a reporting battle. *Nature*, 443, p. 392.

Soete, L., 2016. UNESCO science report (2015) toward 2030. *Round Table Discussion on Monitoring STI for SDG*. United Nation Headquarters.

Tagbo, E., 2010. *Media Coverage of Climate Change in Africa A Case Study of Nigeria and South Africa*. Reuters Institute Fellowship Paper, University of Oxford.

Takahashi, B. and Meisner, M., 2012. Climate change in Peruvian newspapers: The role of foreign voices in a context of vulnerability. *Public Understanding of Science*, 22(4), pp. 427–442.

Tolan, S., 2007. Coverage of climate change in Chinese media. *Human Development Report 2007/2008*. Human Development Report Office Occasional Paper, UNDP.

Tran, M., 2021. *Science journalism for development in Vietnam: A multi-facet investigation of issues and challenges and its implications for Global South*. PhD thesis. Bournemouth University.

Tran, M. and Nguyen, A., 2023. Besieged from all sides: impediments to science journalism in a developing country and their global implications. *JCOM*, 22(04). https://doi.org/10.22323/2.22040204

UNESCO (2010) UNESCO Science Report 2010. Poland: UNESCO.

Verghese, S., 2016. A comparative analysis of the news coverage and content of science and technology in two English dailies. *Anveshana's International Journal of Research in Regional Studies, Law, Social Sciences, Journalism and Management Practices*, 1(11), pp. 78–79. https://ssrn.com/abstract=2896870

Víctor, A.M., 2019. *Reporting science journalism in Peru: A qualitative study of challenges and opportunities*. MA thesis, Hochschule Bonn-Rhein-Sieg.

Williams, K., 2011. *International Journalism*. Thousand Oaks: Sage.

7

"THE DEATH OF EXPERTS"

Sourcing science journalism in a state-controlled media environment

Alexandra Borissova Saleh

Talking to people is one of the cornerstones of journalistic work. Indeed, journalism might be summed up as passing one person's words on to another. In countries where the state controls – or attempts to control – public discourse, talking to people can be both difficult and dangerous. A key means by which an authoritarian state can exert control on journalists is by limiting their access to sources. When a state is in crisis, whether due to natural or political events, access to sources is likely to be even more severely curtailed. Taking Russia as an example of a media environment with high levels of state control, this chapter examines the impact of crisis conditions on science journalists' interactions with their sources and the politicisation of the beat that this entails.

As discussed in Chapters 4 and 5, science journalists rely heavily on written materials such as peer-reviewed papers, press releases and other institutional communications. But if the journalist is to understand the context of these written pronouncements, to produce reports that appear credible or to add colour and new angles to their reporting, they also need to talk to people – including researchers and scholars, industry experts and policymakers, as well as ordinary people whose lives might be affected by the activities (or inactivity) of the former. As American science journalist Melinda Wenner Moyer notes: "In journalism – and science journalism in particular – success is all about who you talk to" (Moyer, 2020: p. 27).

Yet science journalists' interactions with their sources are inherently challenging, in part because time flows differently in the worlds of the media and scientific research. Journalists look for well-defined discrete events that occur in a short period and are high in news values (see Chapter 9), whereas researchers experience science as a gradual incremental process typically made up of small tentative steps that may not have high significance in themselves.

DOI: 10.4324/9781003326724-10

Routine science journalism connects these two contrasting timescales through the convention of treating the publication of peer-reviewed papers as potential news events. This is accompanied by a shift away from the scientists' emphasis on method, which configures the published study as one step in the ongoing activities of the wider research community, to a focus on the application and relevance of the study – a move that implicitly configures the results reported as certain and complete. This conventional reliance on peer-reviewed papers directs journalists towards preferred sources – the scientists who authored the paper. While some journalists may also seek out additional sources who were not involved in the study for an independent view, by relying on peer review to generate news events, science journalists routinely delegate judgement of the credibility of sources to the peer reviewers – the experts working in the research field whom the scientific journal has invited to judge whether a paper is methodologically rigorous, properly executed and generating sufficiently significant results to merit publication.

In crisis conditions, this compromise – the delegation of normal journalistic judgement in return for identifiable news events – breaks down. This was the case during the Covid pandemic when scientific findings were fed into public discourse and policymaking much faster than the normal peer-review process would allow, through the circulation of preprints (draft papers that have not yet been peer reviewed) and other direct interventions. In such situations, the burden of judging the credibility of sources and identifying the current scientific consensus falls back on the journalist. At the same time, the significance of potential policy implications means that science news is upgraded from soft news to hard news, bringing all kinds of social and political pressures to bear on both journalists and their sources.

Access to government and state-affiliated experts was crucial during the pandemic. A study of pandemic reporting in the Spanish press found that together government organisations and governmental scientific organisations (such as the National Regulatory Medicines Agency) made up 40% of all the sources used (Catalan-Matamoros and Elías, 2020). Sourcing strategies can also be associated with the politicisation of an issue. For instance, studies of US science reporting of both the Covid pandemic and climate change have found that journalists' increasing use of political, rather than scientific, sources contributed to a politicisation of science news (Chinn et al., 2020; Hart et al., 2020). In media environments with significant levels of state control, such as Russia, the causation may flow in the opposite direction, with heightened political control curtailing journalists' sourcing options.

To explore this dynamic, this chapter draws on interviews with Russian science journalists, who reflect on how the Covid pandemic that started in 2020 has impacted on their work, especially in terms of their interactions with sources (see also Litvinenko et al., 2022).[1] The Russian case also provides a (fortunately rare) example of science journalists working in a country engaged in an aggressive war following the invasion of Ukraine in February 2022. These two successive crises exposed science journalists to politicisation of their work, not only in the newsroom through explicit or implicit editorial guidelines and direct government interference, but also as a result of the changing behaviour of their sources and self-censorship.

From diversification to repression

To understand the context in which Russian science journalism has been politicised, it is first necessary to sketch out the recent history of Russian journalism more generally. The evolving media landscape in Russia can be understood as an example of what Andrew Chadwick (2013) calls a hybrid media system, in which the old media logics of state-controlled broadcasters and newspapers persist, to a varying degree, alongside newer media logics of independent journalism and online news.

After the fall of the communist regime in 1991, the media entered a honeymoon period of post-glasnost journalism with a diverse media landscape representing various owners and political parties. Formal and informal codes of conduct that appeared during this time underlined personal morals and decency, rather than alignment with the Communist Party as previously. This diversity did not last, however. Soon media ownership became concentrated in the hands of oligarchs whose companies published election propaganda and engaged in kompromat wars, a common political tactic in Russia involving the circulation of compromising material.

During Vladimir Putin's first presidential term, TV channels were gradually returned to direct state control (Burrett, 2011) and all traditional media were subjected to highly restrictive regulation. However, the new and strongly growing sector of online media remained independent for some time (Brenton, 2011). Thanks to internet deregulation in the 2000s, online media in Russia were able to develop freely, producing relatively balanced news coverage that gave voice to different parts of the political spectrum (Denisova, 2017). As a result, unlike in other countries, the online media landscape in Russia does not mirror the offline media system (Bodrunova and Litvinenko, 2021). Online-only actors, such as *Gazeta.ru* and *Lenta.ru*, have played – and continue to play – a crucial role in the provision of information of all kinds.

After the wave of protests that began in 2011, the major online media outlets were also put under government control, mostly through a change in ownership (Vendil Pallin, 2017). The new owners, tycoons who were close to the regime, fired whole editorial teams and appointed editors-in-chief who were prized for their loyalty rather than their professionalism (Meduza, 2016). Legislative repression followed. One of the first laws heralding a period of growing internet control was the 2012 amendment to the law on the Protection of Children from Information Harmful to their Health and Development. While ostensibly this law aimed to protect children from dangerous content online, from the beginning media professionals understood it as a tool for internet censorship (Litvinenko and Toepfl, 2019). In the coming years, more restrictive laws followed under the guise of protecting the public against extremism, terrorism and misinformation (Jackson, 2016; Kravchenko, 2019). Online providers working with incoming international traffic were required to install state monitoring equipment, and international platforms were required to store the data of Russian citizens in Russia by physically placing their servers inside its territory. These laws broadened the power of the mass communication watchdog Roskomnadzor, allowing it to act without court rulings (Ermoshina et al., 2022). The subsequent actions against journalists were showcased by state media and contributed to an increase of self-censorship among journalists (Bodrunova et al., 2021).

The government used the Covid pandemic to extend its control, including sentences of up to five years' imprisonment for spreading 'fake news' about the virus (Nadeau, 2020). The beginning of the war in Ukraine brought further restrictions, including the introduction of full-scale repression of the Russian media through a series of measures that effectively criminalised independent journalism. The treason statutes were broadened and protests outlawed, legalising severe punishments for protesters. These are supplemented by illegal measures such as assault and torture (Lokhmutov, 2022). The websites of independent news outlets were blocked within the territory of the Russian Federation by direct order of the General Prosecutor's Office. The list of blocked news sites runs to dozens of titles and includes very different kinds of outlets: from the 1990s flagships of liberal journalism, such as *Echo of Moscow* and Nobel peace prize winner Dmitry Muratov's *Novaya Gazeta*, to small local news websites such as *Bumaga* in Saint Petersburg or *Taiga.info* in Novosibirsk (Takie Dela, 2022). In addition, any kind of Ukrainian news site, foreign broadcasters such as the BBC and Deutsche Welle and major opposition news outlets such as *TV Rain* and *Meduza* were also banned (Troianovski and Safronova, 2022). So, too, was access to Facebook and Instagram; as these services' parent company, Meta, was proclaimed an extremist organisation, journalists had to suspend their Facebook accounts amid fears of being prosecuted.

Many news organisations and their reporters and editors were denounced as foreign agents by the Russian state and foreign correspondents left Russia over security concerns. The last remaining independent news outlets were officially shut down. Some – like *Novaya* and *TV Rain* – later reopened elsewhere in Europe, a move that also poses numerous limitations to their work. Independent journalism thus moved abroad and onto social media channels, especially YouTube and Telegram. Employees of the relocated outlets could not return to Russia, even temporarily, without risking immediate arrest as foreign-based media outlets were designated undesired organisations. All activities of undesired organisations are considered illegal and collaboration with them is criminal. Funding also dried up, with shrinking advertising revenue and few opportunities for crowdfunding. What remains of the Russian media is deeply fragmented and politically polarised, with low levels of public trust (Bodrunova and Litvinenko, 2015; Ipsos, 2021) and a very high level of instrumentalisation by the state (Edelman, 2022).

Science media in post-Soviet Russia

Science journalism has experienced similar forces over the last three decades, but is more strongly rooted in the Soviet past in which many popular science magazines originated. As a result, Soviet-born science magazines coexist with science journalism at the more-or-less independent news outlets and the newer digital-native science magazines (Borissova and Malkov, 2020).

The oldest and most celebrated Russian science magazine, *Nauka i Zhizn* (*Science and Life*), was founded in 1890 and started as a 16-page weekly digest. During the Soviet era, it developed into a monthly 64-page magazine with a circulation of around 3.5 million by the last decades of the Soviet regime (Yakovenko, 2012). In its golden age, the magazine was headed by Rada Adzhubey, daughter of Nikita Khrushchev, the First Secretary of the Communist Party, and was hence closely aligned with the party agenda. This assured its extremely high circulation. Other popular science magazines also enjoyed high circulation: 700,000 copies a month for *Znanie-Sila* (*Knowledge is Power*) and 300,000 for *Khimiya i Zhizn* (*Chemistry and Life*). These, and about a dozen other titles, could be found in almost any household. Although focused on natural sciences, dialectic materialism and scientific communism, such magazines also contained crosswords, cooking recipes and even elements of new-age thinking (which was quite widespread among the Soviet intelligentsia) to provide entertainment for the entire family. What was lacking was any form of investigative journalism, critical views, contemporary social sciences or discussions about science policy.

After the Soviet Union fell and its propaganda machine ceased to exist, science magazines reduced to a very modest market share, becoming a niche product for a narrow science-oriented audience. *Nauka i Zhizn*, for instance, increased in size to a thick 144-page monthly publication, but with a much reduced circulation of 36,000 copies, just 1% of what it had been. As Russian science lost its funding and Russian scientists left the country, popular science also lost its prominence.

However, with the recovery of the Russian economy after the financial crisis of 1998, interest in both science and popular science also recovered. Independent radio stations of all kinds were a prominent feature of the media market in this period and important programmes about science appeared, such as *Granit Nauki* at the liberal, and since 2022 discontinued, radio station Echo of Moscow. Science also began to appear in the general interest media outlets whose independent and competitive approach attracted a quickly growing readership in the 2000s. Political or business magazines such as *SmartMoney*, *The New Times*, *Itogi* and *Russian Newsweek* were the first to be able to afford elaborate science sections, typically produced by reporters with scientific backgrounds. The lead was soon taken by a new type of outlet – digital-native newspapers – which were actively hiring scientists (still badly underpaid as researchers) to work as specialist science reporters. Three key players in this market were *Gazeta.ru*, *Lenta.ru* and *Polit.ru*, all founded as online-only news outlets in 1998 and 1999 (Sapun, 2016). These outlets became a gateway to the profession and a school for a number of reporters who later created independent projects that continue to shape Russian science journalism.

From 2013, as nearly all independent media were put under some form of governmental control, many prominent science journalists left the general-interest news outlets, along with colleagues who covered economics, politics and social issues. Yet the Russian public's interest in science endured (Vaganov, 2016), so a number of journalists established independent popular science websites that largely replaced the Soviet-era science magazines as sources of scientific information. They did not, however, mimic the older print magazines in serving as an instrument of state propaganda; their niche topic allowed them, for the time being, to maintain relative independence. The most read of these digital magazines, including *Popular Mechanics*, *National Geographic Russia* and *N+1*, enjoy monthly audiences of several million readers. They have thrived, in part, due to the lack of competition from traditional news outlets, where popular science was deliberately replaced by para-scientific stories about astrology, conspiracy theories and other kinds of pseudo-science.

The new digital science magazines were technologically advanced and appealed to younger audiences through their use of mobile apps, close integration with social media and their reporters' engagement with the global science journalism community. However, this success has entailed some compromises. First, to avoid the fate of the general interest outlets, the online magazines have focused on research news rather than science policy. Second, they largely rely on state funding, either directly from the Ministry of Science and Education or indirectly from leading Russian universities and research centres, which are themselves dependent on government support or that of largely state-owned high-tech industries. To avoid the media repression that came with the start of the war in Ukraine, popular science magazines typically avoided topics of social significance. None were shut down or blocked like the general interest outlets, but they suffer from the same legal limitations as other journalists. Those science journalists who have not emigrated have had to suspend their social media pages to be able to continue working with government-funded sponsors. The risk of relying on state-funding also became apparent, as funding timescales shrank to just two or three months. As the editor of one of the magazines puts it: "There is no direct pressure on us, we will simply starve to death. Before we were eating from the recycle bin, and now there is nothing even there."

The question, then, is whether independent science journalism still exists in Russia. Russian journalists draw an analogy with the double white lines in the middle of a road (Gehlbach et al., 2023; Yablokov, 2016): although non-state-owned media might not always be subject to direct censorship, they are free to express themselves only within bounds, the double white lines that must not be crossed. While for the news journalist, the double white lines may mean avoiding criticism of Vladimir Putin, for the science journalist, they mean avoiding political issues, itself a form of politicisation. As will be seen below, the constant awareness of these double white lines makes self-censorship an ever-present feature of the culture of Russian science journalism.

Science journalism was forced to confront the double white lines of Russian journalism during the Covid pandemic, when science reporting was abruptly upgraded from soft news to hard news. The emergency powers that the Russian government gave itself to suppress dissent meant that every topic of social importance – including Covid itself – became political, attracting the scrutiny of all kinds of controlling bodies. This put an unusual pressure on science journalists, both at pro-state outlets and at the remaining islands of independent reporting; pressure that was further intensified by the war in Ukraine. One of the most significant impacts was on journalists' interactions with sources, most of whom were themselves employed by state-owned institutions and thus also working under severe constraints.

The following discussion draws on interviews conducted in 2022 with 23 science journalists who work at a variety of Russian media outlets. Their testimonies reveal that the journalists' selection of sources, their access to sources and their judgements about source credibility were all affected by the politically charged environment, though exactly how depended on whether the journalist worked at an independent or a pro-state outlet.

Source selection

Globally, the pandemic presented a challenge for journalists as it became essential to gather and distribute scientific information to the public, requiring extensive interactions with a range of experts, including doctors, virologists, epidemiologists, public health experts and policymakers (Perreault et al., 2022). Russia was no exception here. The pandemic raised the profile of science journalism in the country. With more resources and time, science journalists had the opportunity to reach out to more experts and institutions. In their interviews, the journalists at specialist science outlets recall interacting with more sources. In some cases, this has led to strong lasting connections with research institutes and medical universities. A wider range of experts was also called on; for instance, clinical practitioners who had previously been viewed as "not scientific enough" were now considered useful sources.

Difficulties gaining access to sources within Russia led some popular science outlets to focus solely on sources based abroad. However, this did not necessarily mean neglecting Russian scientists. Repeated waves of emigration have led to a large diaspora of scientists with good academic credentials who still speak Russian and who remain connected with their homeland (Belousova et al., 2012; Puffer et al., 2018). The Russian-American Science Association helped connect these experts, based all over the world, with Russian journalists through a Facebook group where they could interact with each other directly.

In contrast to the expanded range of sources used by the journalists at the specialist science outlets, science reporters at general interest and state-owned outlets opted to stick to their established verified contacts. Interacting with people who already knew them enabled them to get the latest information, sometimes unofficially. One reporter at a state news agency recalls: "When the first woman died [of Covid], I had an interview with Dr X, and he talked to me directly, without passing through the press office. He described in detail the progress of the disease and the treatment – in accordance with personal data protection but without trying to hide anything." Interacting with familiar experts also allowed these journalists to get comments much faster, even within an hour, which was particularly valuable under the extreme time pressure of the first months of the pandemic. Experts of all kinds, from researchers and doctors to state functionaries at regional and federal levels, were available to these journalists for rapid responses.

With the start of the war, many of the independent outlets were forced abroad. The journalists in exile were limited to interacting with their (often anonymous) sources online or monitoring social media posts. This made normal interactions with sources impossible, as a reporter at a general-interest opposition news outlet explains: "You cannot reason on the situation only through social media, you have to be there on the spot and speak to people directly."

Covid lockdowns meant journalists around the world had experienced some similar logistical limitations. TV reporting, in particular, needed to work around the inability to conduct face-to-face interviews. However, Russian journalists in exile faced additional challenges, as the reliance on online communication brought with it the risk of government surveillance. Journalists therefore switched to protected means of communication, like the encrypted messaging app Signal. Yet this also posed difficulties. Even directing someone towards Signal may appear suspicious, as one independent journalist notes:

> I cannot write to a source at their university mail address 'let's talk about state treason in Russia', because I have no doubt who is reading this mailbox. But when I look for a person, the son of a prisoner who is an active scientist, I only find the university mail address. And somehow, I need to tell him 'write to me on Signal'. It is a danger on both sides, the scientist and the journalist are at risk over the article.

With the war, the differential selection of sources already present during the pandemic became complete, leading to the formation of social bubbles that barely communicate with one another. Both journalists and experts now self-select to speak only to people with like views. According to an editor of a digital news magazine, this is the case even when selecting sources for science stories unrelated to the war itself:

> After the start of the war, we do not really want to talk to those who openly support the war. This is not an editorial order – but this is the selection bias that guides me and my colleagues. It's not a rule, it's an informal practice that reflects our own opinion about the war.

Source availability

The use of sources depends not only on their selection by the journalist but also on whether the source is prepared – or permitted – to speak to the media. As most scientists in Russia are employed by public (i.e., state-controlled) institutions, they have experienced the same controls on their interactions with the media as government officials experience. With the outbreak of the pandemic, managers at public institutions quickly banned any unapproved contact with journalists – customary practice for government officials – and obliged everyone to contact their public relations offices first. Furthermore, the key aspects of the pandemic response (PCR testing, statistics, vaccination) were entrusted to employees of the Ministry of Health or the federal consumer protection and wellbeing agency, Rospotrebnadzor (RPN). Both of these organisations have restrictive information policies whereby external communication is only allowed with prior approval. Communicating with scientific experts at these institutions is effectively the same as communicating with civil servants or policymakers, already difficult for independent news outlets even before the pandemic.

As a result, journalists working at independent and state media in Russia report diametrically opposed experiences of the availability of sources inside Russia during the pandemic. State-owned outlets were granted easy access to the most valuable sources, such as the creators of the first PCR tests at the semi-military Vektor Institute in Novosibirsk and the developers of the Sputnik vaccine at the Gamaleya Institute in Moscow. Journalists at these outlets found that both officials and scientists were eager to talk to them:

> Our scientists were always happy to answer our questions. Their communication was timely and clear, interaction through press offices was very comfortable. There was absolutely no position: 'You don't know anything, what to talk to you about?' I want to say thank you, it was the best interactions with scientists through my whole long career.

The experience of independent journalists was drastically different. These journalists experienced major problems contacting both official representatives and Russian scientists, in part due to gatekeeping by their PR officers but also due to self-censorship on the part of the sources. Several types of obstacles were encountered. One was 'editorial volleyball', where a request for an interview is forwarded from one official or organisation to another until it is no longer timely. For example, the editor of a digital science magazine recalls:

> The first time I called Gamaleya Institute was at the beginning of summer. They sent me to the Ministry of Health; it would comment in a week. I bombarded them with calls and email for 64 days in a row. They answered me politely, but I did not got any real answer to any of my questions.

Other delaying tactics, such as requiring that the request be sent on headed paper, had a similar effect. In other cases, some journalists' requests – especially those from smaller local news outlets – were ignored completely, even while the same government sources organised regular press conferences for foreign journalists: "Together with colleagues from other regions we requested several times an interview with Vektor representatives, but RPN did not respond to our requests. The pool of experts they provide was inaccessible."

A science journalist at a digital magazine sums up the situation:

> For half a year we received no response from either the Ministry of Health or RPN. At the same time, I cannot say that there was no information for the press. They gave interviews, participated in press conferences. But the media that was allowed to talk to the developers [of the vaccine] is a narrow circle.

With the beginning of the war in Ukraine, the availability of government sources in Russia was blocked entirely for independent science journalists.

The war made reporting science seem like a low priority for both journalists and their sources. For instance, an editor of a digital science magazine recalls: "Especially in the first month after the start of the war, there were several cases when a source said, I am too sad because of the war, I don't really want to talk about science in general now." Even regular sources were reluctant to engage with journalists during this time. "All of us found ourselves in a situation where you do not understand what is the point of doing what you are doing", the same editor continues. "It is possible, but who needs it when people are dying in the Ukraine?" At times, a similar sentiment led to journalists, too, adapting their behaviour: "self-censorship has appeared, that you feel bad about yourself, because you are Russian. Many people in our editorial office are profoundly worried about the war, and not in the sense 'oh, when it is that we take Kiev?', you know what I mean?"

Faced with limited, and eventually no, access to Russian experts, an alternative was for journalists to turn to sources outside Russia, as noted above. Almost all the journalists interviewed report that during the pandemic, international experts were more open and clearer in their communication than researchers working in Russia. For instance, a reporter at a digital science magazine observes:

> Many government agencies or scientists who are associated with them work with near censorship – their comments must be checked and agreed upon at dozens of levels before being given to us. Often scientists from abroad respond more easily and faster, as in their picture of the world, science relies on taxpayers' money and they feel obliged to respond, but in Russia this is not always the case.

The war in Ukraine complicated the relationship between Russian journalists and international sources, some of whom boycotted contact with Russian media. However, science reporting did not seem to suffer from this 'cancel culture', perhaps due to the cosmopolitan nature of both science journalism and science itself. As the editor of a digital science magazine puts it: "We do not have any issues with our requests sent abroad. No one answered, like, f*ck off, go shoot Putin – no, this was not the case." The willingness of Russian scientists based abroad to serve as independent, transparent and accessible expert sources suggests that the lack of openness at Russian institutions cannot be attributed to some cultural peculiarity on the part of Russian scientists. Rather, it is the consequence of the structure and regulation of Russian science. Scientists who remain in Russia need great courage to speak openly. Rather than test the limits of the system, many decide to self-censor by avoiding media contact altogether.

Source credibility

Public trust is a crucial factor for any society when it faces a crisis (Phillips, 2021). The effect of the pandemic on different societies was strongly dependent on the previous history of trust – in the media, in institutions, in science and medicine – in that society. Non-democratic societies, including Russia (Shlapentokh, 2006) and China (White and Fu, 2012), are traditionally characterised by low levels of institutional trust and trust in the media. This lack of institutional trust manifests itself in the relationship between journalists and their sources, especially official ones.

All the respondents commented on their decreasing trust in sources during the pandemic. Representatives of governmental bodies were considered the least reliable. In contrast to the Centers for Disease Control and Prevention in the United States or the European Medicines Agency in Europe, neither RPN nor the Ministry of Health were seen as a credible source of data by journalists. Rather, these organisations added to the work of science journalists by spreading various kinds of misinformation that journalists needed to debunk themselves in the absence of independent experts able, and willing, to speak out. A science journalist at a pro-state general interest newspaper cites the example of using gloves: "Both RPN and the Ministry sometimes sent so much inadequate information that we were trying to write back to them to explain their mistake in a diplomatic way."

Eventually, science journalists came to use the official sources only to report the official position or new regulations (such as lockdown measures) rather than to provide any explanation or guidance, for fear the latter would reflect party policy rather than the actual situation. Yet outlets that did not have a dedicated science section did pass on RPN statements unquestioned. A science journalist from an oppositional outlet blames this for the chaos that characterised the pandemic response in Russia: "You go to a simple postdoc, and they help sort out the nonsense that official experts have said.... Many Russian media were responsible for that hell."

Faced with the uncertainty of the situation and the inevitable disagreements about the best course of action, journalistic judgements of the reliability, up-to-dateness and impartiality of sources were more important than in normal times (Pentzold et al., 2021). Science journalists therefore became more rigorous in exercising their own judgements about the claims their sources made. As the editor of one digital science magazine explained: "About any other disease, you see a consensus and it is more or less the same, but here you had to look at everything. The circle of sources has expanded, but the circle of experts has narrowed, and the degree of trust in everyone has fallen."

According to a reporter from a state news agency, the pandemic exposed the weaknesses of the Russian scientific community:

> Narrow experts are needed. It is difficult to find epidemiologists, they do not exist. They disguise themselves as mathematicians, virologists, biologists. They refused to comment. And those handful who are reasonable are so exhausted that at some point they turn off the phone. There are no certainly reliable sources. We are in an extraordinary position; we can watch how scientists study the epidemic in real time.

Overly narrow expertise was also a problem, as a digital science magazine editor notes:

> What I did not expect is that our medical community are often so immersed in their narrow specialisation that on other topics they talk the same nonsense as the layman. [Some doctors] are good in their field, but no longer open scientific journals since they graduated from the university. I realized that I needed to be much more careful than I thought.

For one newspaper journalist, this over-extension of expertise amounted to pseudoscience: "Pseudoscience within science has spread. When someone tries to cling to research or expresses an expert opinion that is not at all from his sphere, so he is incompetent."

In democratic countries, transparent journalism is held up as the ideal, with named sources being assumed to be more credible than anonymous sources. For journalists working under the constraints of an authoritarian regime, anonymity is often unavoidable. Nearly all science reporting concerning the war – such as comments on Russian Defence Ministry statements about secret labs in Ukraine – was anonymous. Co-operation between Russian and Ukrainian scientists, geography, everything that has to do with the territory: all became too threatening for the personal security of named sources. For one independent journalist, the consequent reliance on anonymous sources overturns the very notion of an expert: "It is a triumph of anonymity and the death of experts. If an expert gives an opinion anonymously, what's the point? We consider an expert an expert, because they have a surname and a track record and we trust them. Otherwise… ."

This epidemic of anonymity also infected journalists themselves. Those who have stayed in Russia and still have a platform to express themselves openly (for instance, by freelancing for international media outlets) must do so anonymously. One independent science journalist says they decided to write anonymously while still in Russia simply to call the war a war. Such journalists must choose between three evils: writing anonymously, self-censoring, or being prosecuted for spreading 'fakes' and even for treason if working for a foreign publisher. Clearly, the first of these is the most preferable option.

The combination of severe media restrictions and crisis conditions makes it difficult to assess the credibility of sources. Journalists must spend more time assessing whether a source is reliable and trustworthy and whether they have the appropriate expertise. They must then also assess whether it is safe to name the source, and whether to name themselves – two decisions that risk further undermining the credibility of both the source and the journalism.

Conclusion: The one-two punch against Russian science journalism

Journalism consists of talking to people and then telling others what they have said. In a state-controlled media system, talking to people can be hard and it becomes even harder in crisis conditions or when government policy is at stake. This is what Russian journalists have experienced in recent years, first during the pandemic and then during the war in Ukraine. One journalist aptly sums up the situation with a boxing metaphor:

> I describe our situation as a one-two punch, you know, in boxing when you get the second punch before somehow managing to recover from the first one. Starting with the pandemic, we weren't given any comments anyway. That made our job harder. And now we are being deprived of advertisers and sponsors or simply shut down.

When science journalists report on science-based policy stories and stories of social importance, they need to be able to talk to scientific experts, who in turn need to be able to make public statements about politicised issues. In any media system, the use of expert sources can be narrowed by journalists' reliance on familiar contacts. However, in authoritarian regimes, access to experts is further limited – at best controlled and at worst confined to pro-government messaging. Of course, even experts free of government constraints are influenced by their own paradigms, cognitive biases and cultural and political values, but nevertheless they represent a crucial element of the ecosystem that allows a good journalist to create a balanced picture for their readers. In state-controlled media systems such as Russia, this balance can easily be blocked.

Another problem is the personal safety of journalists and sources, fears over which push both towards anonymity. To have properly functioning public expertise, sources (and the journalists who talk to them) need to be named so that their credibility can be judged. They also need to be able to speak openly so that it is their expert opinion, not the government position, that is heard. Where these conditions are not possible, public expertise cannot exist.

The extinction of public expertise further inhibits independent science journalism. With the first punch delivered by the Covid pandemic, the public dissemination of credible expertise was vital. Yet the unknown virus and the uncertain consequences of policy responses meant that the usual means of judging scientific credibility were too slow. As a result, journalists needed to focus on their role as information brokers, judging for themselves who was credible and who was not; frequently, Russian science journalists concluded that government officials were not.

After the war delivered the second punch to the Russian media, science journalists were effectively deprived of their purpose, as they lost the ability to play the role of a detached mediator and were forced instead to take up a position on one of the poles of an increasingly polarised society. Those who worked in the state-owned media effectively became propagandists, since sources of government propaganda were the only kind of sources they were allowed to use. Journalists at independent outlets, on the other hand, found themselves playing the role of activists rather than journalists. The phenomenon of general news journalists turning into activists has been observed in other restricted media markets, including investigative journalism in China (Wang, 2016), the online media of the Ethiopian diaspora (Skjerdal, 2011) and Bosnian journalism during the war with Serbia (Ruigrok, 2010). In the case of Russia, the science beat has also faced this dilemma when first Covid and then the war in Ukraine politicised science journalism. With public expertise an impossibility and journalists no longer able to function as mediators of different viewpoints, Russian science journalists have been forced to choose: do they serve as state propagandists or as activists?

Note

1 If not otherwise referenced, quotes come from these interviews.

References

Belousova, M., Ataiants, J. and Olimpieva, I., 2012. The Russian diaspora in the USA. *Russian Analytical Digest*, 107, pp. 2–9.

Bodrunova, S.S. and Litvinenko, A.A., 2015. Four Russias in communication: Fragmentation of the Russian public sphere in the 2010s. In: Dobek-Ostrowska, B. and Glowacki, M. (eds.), *Democracy and Media in Central and Eastern Europe: 25 Years On*. New York: Peter Lang, pp. 63–79.

Bodrunova, S.S. and Litvinenko, A.A., 2021. Media accountability in a polarized society. In: Fengler, S., Eberwein, T. and Karmasin, M. (eds.), *The Global Handbook of Media Accountability*. Abingdon: Routledge, pp. 63–79.

Bodrunova, S.S., Litvinenko, A. and Nigmatullina, K., 2021. Who is the censor? Self-censorship of Russian journalists in professional routines and social networking. *Journalism*, 22(12), pp. 2919–2937.

Borissova, A.O. and Malkov, D.V., 2020. Russian pendulum: From glorious science propaganda to modest public engagement initiatives. In: Gascoigne, T. et al. (eds.), *Communicating Science: A Global Perspective*. Acton: ANU University Press, pp. 715–741.

Brenton, T., 2011. Russia's media: Freedom isn't dead. *British Journalism Review*, 22(1), pp. 33–39.

Burrett, T., 2011. *Television and Presidential Power in Putin's Russia*. Abingdon: Routledge.

Catalan-Matamoros, D., and Elías, C., 2020. Vaccine hesitancy in the age of coronavirus and fake news: Analysis of journalistic sources in the Spanish quality press. *International Journal of Environmental Research and Public Health*, 17(21), pp. 8136–8150.

Chadwick, A., 2013. *The Hybrid Media System. Politics and Power*. Oxford: Oxford University Press.

Chinn, S., Hart, P.S. and Soroka, S., 2020. Politicization and polarization in climate change news content, 1985–2017. *Science Communication*, 42(1), pp. 112–129.

Denisova, A., 2017. Democracy, protest and public sphere in Russia after the 2011–2012 anti-government protests: Digital media at stake. *Media, Culture & Society*, 39(7), pp. 976–994.

Edelman, R. 2022. Truth will out; or will it. *Edelman*, 24 Aug., https://www.edelman.com/insights/truth-will-out-or-will-it-6-am.

Ermoshina, K., Loveluck, B. and Musiani, F., 2022. A market of black boxes: The political economy of internet surveillance and censorship in Russia. *Journal of Information Technology & Politics*, 19(1), pp. 18–33.

Gehlbach, S., Lokot, T. and Shirikov, A., 2023. The Russian media. In: Wengle, S. (ed.), *Russian Politics Today*. Cambridge: Cambridge University Press, pp. 390–407.

Hart, P.S., Chinn, S. and Soroka, S., 2020. Politicization and polarization in COVID-19 news coverage. *Science Communication*, 42(5), pp. 679–697.

Ipsos, 2021. *Global Trustworthiness Index*. Ipsos, https://www.ipsos.com/sites/default/files/ct/news/documents/2021-10/Global-trustworthiness-index-2021.pdf.

Jackson, C., 2016. Legislation as an indicator of free press in Russia. *Problems of Post-Communism*, 63(5-6), pp. 354–366.

Kravchenko, M., 2019. Russian anti-extremism legislation and internet censorship. *The Soviet and Post-Soviet Review*, 46(2), pp. 158–186.

Litvinenko, A., Borissova, A. and Smoliarova, A., 2022. Politicization of science journalism: How Russian journalists covered the covid-19 pandemic. *Journalism Studies*, 23(5-6), pp. 687–702.

Litvinenko, A. and Toepfl, F., 2019. The "gardening" of an authoritarian public at large: How Russia's ruling elites transformed the country's media landscape after the 2011/12 protests "for fair elections". *Publizistik*, 64, pp. 225–240.

Lokhmutov, A., 2022. Russian police are torturing anti-war activists. *Human Rights Watch*, 20 Oct., https://www.hrw.org/news/2022/10/20/russian-police-are-torturing-anti-war-activists.

Meduza, 2016. 12 redakzy za pyat let Razgony, blokirovki, uvolnenia glavredov: kak vlast reformirovala rynok SMI [12 editorial offices in five years: Crackdowns, blockings, dismissals of editors-in-chief: how government reformed the mass media market]. *Meduza*, 17 Mar., https://meduza.io/feature/2016/05/17/12-redaktsiy-za-pyat-let.

Moyer, M.W., 2020. Sources and experts: Where to find them and how to vet them. In: Blum, D., Hatch, J. and Jackson, N. (eds.), *KSJ Science Editing Handbook*. Cambridge: Massachusetts Institute of Technology, pp. 27–38.

Nadeau, J., 2020. Putin is using the pandemic to consolidate power. *Foreign Policy*, 18 May, https://foreignpolicy.com/2020/05/18/putin-is-using-the-pandemic-to-consolidate-power/.

Pentzold, C., Fechner, D.J., and Zuber, C., 2021. "Flatten the curve": Data-driven projections and the journalistic brokering of knowledge during the COVID-19 crisis. *Digital Journalism*, 9(9), pp. 1367–1390.

Perreault, G., Perreault, M.F., and Maares, P., 2022. Metajournalistic discourse as a stabilizer within the journalistic field: Journalistic practice in the Covid-19 pandemic. *Journalism Practice*, 16(2-3), pp. 365–383.

Phillips, C. 2021. *Ipsos Global Trustworthiness Monitor: Is Trust in Crisis?* Ipsos, https://www.ipsos.com/sites/default/files/ct/news/documents/2022-01/Global-Trustworthiness-Monitor-2021.pdf.

Puffer, S.M., MacCarthy, D.J. and Satinsky, D.M., 2018. *Hammer and Silicon: The Soviet Diaspora in the US Innovation Economy*. Cambridge: Cambridge University Press.

Ruigrok, N., 2010. From journalism of activism towards journalism of accountability. *International Communication Gazette*, 72(1), pp. 85–90.

Sapun, A., 2016. Istoria razvitia internet-SMI v Rossiye [History of online mass media in Russia], *DNI.ru*, https://dni.ru/academ/2016/8/1/347408.html.

Shlapentokh, V., 2006. Trust in public institutions in Russia: The lowest in the world. *Communist and Post-communist Studies*, 39(2), pp. 153–174.

Skjerdal, T.S., 2011. Journalists or activists? Self-identity in the Ethiopian diaspora online community. *Journalism*, 12(6), pp. 727–744.

Takie Dela, 2022. SMI, sablokirovannye i zakrytye v Rossii posle nachala "voennoi specoperatsii" na Ukraine [Mass media blocked in Russia after the beginning of the "special military operation" in the Ukraine]. *Takie dela*, https://takiedela.ru/list/zablokirovannye-smi/

Troianovski, A. and Safronova, V., 2022. Russia takes censorship to new extremes, stifling war coverage. *New York Times*, 4 Mar., https://www.nytimes.com/2022/03/04/world/europe/russia-censorship-media-crackdown.html.

Vaganov, A., 2016. Evolution of forms of science popularisation in Russia: XVIII-XXI centuries. *Science, Innovation, Education*, 3(11), pp. 64–77.

Vendil Pallin, C., 2017. Internet control through ownership: The case of Russia. *Post-Soviet Affairs*, 33(1), pp. 16–33.

Wang, H., 2016. *The Transformation of Investigative Journalism in China: From Journalists to Activists*. Lanham: Lexington Books.

White, J.D. and Fu, K.-W., 2012. Who do you trust? Comparing people-centered communications in disaster situations in the United States and China. *Journal of Comparative Policy Analysis: Research and Practice*, 14(2), pp. 126–142.

Yablokov, I., 2016. Russian journalism's double white lines. *OpenDemocracy Russia*, 22 Jul., https://www.opendemocracy.net/en/odr/russian-media-s-double-white-lines.
Yakovenko, I., 2012. Rynok nauchno-populyarnich jurnalov: Analitichesky obzor [The market of popular science magazines: An analytical review]. *Nazionalnaya Tirazhnaya Sluzhba*, 30 Aug.; reproduced at *AdIndex*, 27 Sep., https://adindex.ru/publication/analitics/channels/2012/09/27/93552.phtml.

PART III

The journalistic voice

8

PARTIAL TO BEING IMPARTIAL?

Debates about balance in science journalism

Felicity Mellor

The norm of impartiality has been one of the most contentious aspects of science journalism. It commits science journalists to ensuring their reporting is not overly influenced by the perspective of just one section of society. In practice, this often means journalists seek out contrasting views when reporting on matters of public policy or on topics that are politically controversial. Throughout its existence, the BBC, the United Kingdom's national public service broadcaster, has been committed to impartiality, to the extent that the principle is fundamentally bound up in its identity. The BBC thus provides an instructive example of the challenges that impartiality can pose when reporting science-based stories, having been uniquely reflexive about how impartiality can be applied and how well it succeeds at meeting its own standards. Together with its global influence, this makes the BBC a revealing case for examining not only how impartiality is applied to science news in practice but also how it is conceptualised by news producers, news sources and news audiences. The impartiality of the corporation's coverage of science has been the subject of formal review and the balance of its climate change reporting, in particular, has repeatedly been the subject of critical commentary. This chapter examines this meta-discourse about impartiality for what it reveals about how journalists and scientists talk past each other when it comes to impartial news reporting. As I discuss, the disagreements are exacerbated by conceptual unclarity over the distinction between impartiality and objectivity and by the inherent difficulties entailed in separating fact from opinion.

The BBC is not alone in having struggled with how – and whether – to apply impartiality to the reporting of science-based issues, nor in being criticised for the resulting coverage. News outlets in many countries have been accused of applying 'false balance' (also sometimes referred to as 'false equivalence') by

DOI: 10.4324/9781003326724-12

giving undue emphasis to views that are counter to the scientific consensus. As early as 1965, Howard Smith, a correspondent for the American Broadcasting Company, identified "false balance" as a problem in television journalism, citing as an example a CBS documentary on the risks of smoking in which doctors who recognised that smoking caused cancer were balanced with representatives of the tobacco industry (Smith, 1965). In recent years, it has more often been campaign groups – anti-vaxxers, creationists, climate deniers and others characterised as 'anti-science' contrarians – who have been identified as benefiting from journalists' use of false balance. Ideological and corporate actors understand that they can most effectively undermine the case for policy action by supporting lobby groups to engage with the science within a policy debate. As the tobacco industry explicitly recognised, emphasising the uncertainties that are present in the relevant science can be sufficient to undermine the case for immediate action (Michaels, 2008; Mooney, 2010; Oreskes and Conway, 2010). Lobby groups thus hope to exploit the media access that journalistic balance can offer by taking the opportunity to make claims about the science.

A number of empirical studies have examined which viewpoints are present in news coverage of science-based issues in order to establish whether such groups do succeed in gaining media access through balanced reporting. Many of these studies have focused on the coverage of climate change, including a series of papers by Maxwell Boykoff assessing the proportion of US and UK news reports that present views that are contrary to the scientific consensus (Boykoff, 2007; 2008; Boykoff and Boykoff, 2004; Boykoff and Farrell, 2019; Boykoff and Mansfield, 2008; McAllister et al., 2021). Such studies find that in some types of coverage in some outlets at some times, climate sceptic views have been relatively common; in other outlets at other times, they are rarely given the same weight as the scientific consensus. Boykoff and his co-authors find an increase in the inclusion of sceptic voices in the US media during the 1990s and 2010s and attribute this to the activities of well-organised sceptic lobby groups who use tactics akin to those developed by the tobacco industry (Boykoff and Boykoff, 2004; Boykoff and Farrell, 2019). However, sceptics have been found to be less common in climate coverage in non-anglophone countries (Dayrell and Urry, 2015; Painter, 2011). Perhaps the most important conclusion to draw from this large body of work is that it is hard to generalise about the prevalence of the practice.

Despite that, individual instances of climate reporting have provoked heated debate about news outlets' use of balance. Critics complain that the practice promotes fringe ideas, misleads the public, delays policy action and discourages behaviour change. How can it be that journalists and scientists – two professions that ostensibly share the aim to uncover the truth – can disagree so consistently and for so long about how science should be reported? This chapter seeks to explore this question by examining the case of the BBC's climate coverage.

Objectivity and impartiality: A short history

First, I consider what is meant by impartiality and the related, but distinct, concept of objectivity. Discussions often conflate these two terms, jumping between the two as if they were synonyms. Other terms also frequently come up: neutral, balanced, disengaged, dispassionate, non-partisan, independent, unbiased, disinterested, detached, fair, accurate, truthful, factual and so on. This excess of vocabulary immediately indicates slippery concepts that are hard to pin down – a conceptual confusion that contributes to the underlying tensions over the reporting of science-based controversies. As I discuss later, these are exacerbated by science's own stake in some of these concepts: science journalists report on claims that have already been assessed for their accuracy and lack of bias.

Ever since news journalism first emerged in its professionalised form more than a century ago, it has been attended in western democracies by the twin aspirations of objectivity and impartiality. In the preceding century, news outlets seeking to combine public interest journalism with a wide market reach unlimited by partisan allegiances had developed a style of reporting that appeared to be politically neutral. In mid-nineteenth-century America, for instance, the penny press attempted to reach a mass audience by stressing its independence from government and political parties (Allan, 1997). This political neutrality was achieved through an editorial approach that emphasised facts. Despite charges of sensationalism, editors claimed they were meeting a public demand for information that was stripped of editorialising explanation and evaluation. As the *New York Herald* put it when it launched in 1835, the aim was to "record facts, on every public and proper subject, stripped of verbiage and coloring" (Allan, 1997: p. 305). Similarly, when Adolph Ochs bought up the *New York Times* in 1896, he immediately announced that the paper would "give the news impartially, without fear or favor, regardless of party, sect, or interests involved" (Ochs, 1896; see Chapter 1). As Ochs made clear when he went on to state his commitment to the causes of tax reform and low taxes, this impartial approach to its reporting did not mean that the paper or its proprietor would forsake all opinion.

Sociologist Michael Schudson argues that these early moves towards fact-based reporting and political neutrality developed into the modern journalistic norm of objectivity in the years after World War I (Schudson, 2001). Schudson suggests that objectivity should be understood as an institutionalised method – akin to the scientific method – for ensuring that journalists adopted a detached and disinterested approach to news. At the heart of this method were two techniques. One was the recently established practice of interviewing news actors and attributing quotes to them; this served as a means of distancing news outlets from the opinions they presented. The other was a clear separation between news reporting and comment.

Newspapers would continue to print comment pieces, but these would be clearly labelled as such. As Schudson puts it: "The belief in objectivity is a faith in 'facts', a distrust of 'values', and a commitment to their segregation" (Schudson, 1981: p. 6). For science journalism, this often meant reporting new research findings as if they were established facts, hastening science along the path to a certainty that is not there in the laboratory (Collins, 1987; Latour and Woolgar, 1986; Nelkin, 1987).

While US newspapers adopted the ideal of objective reporting as a means of demonstrating their lack of partisanship, newspapers in other countries retained their political allegiances. In the United Kingdom, it was rather the development of broadcasting that led to an expectation of editorial neutrality, as the independent BBC needed to make a case that its monopoly of the airwaves would not be a threat to the government if it was allowed to cover controversial news (Scannell and Cardiff, 1991). It did this through a strong emphasis on impartiality.

Having begun as an implicit professional norm adopted by broadcasters for their own purposes, over the course of the twentieth century, impartiality was codified and mandated, becoming a legal requirement for news broadcasters in many countries, especially in Europe (Cushion, 2012: p. 33). In the United States, too, for a time, broadcasting was subject to the Fairness Doctrine, which required broadcasters to ensure balanced coverage on matters of public importance. Today in the United Kingdom, all broadcasters are required by the Communications Act 2003 to present news with due impartiality (Ofcom, 2021). Even where impartiality is not imposed through legislation, public service broadcasters around the world espouse impartiality as an ideal. Both objectivity and impartiality have been enshrined in the ethical codes and editorial guidelines of professional bodies and regulatory authorities (Hanitzsch et al., 2011).

Guidelines on how to produce objective or impartial journalism also often highlight the need for a balanced account of opposing views. For instance, the code of ethics drawn up by the international Ethical Journalism Network notes under the heading of impartiality that: "Most stories have at least two sides. While there is no obligation to present every side in every piece, stories should be balanced and add context" (EJN, 2022). It is this practice that has attracted criticism in the context of science journalism.

The norms of objectivity and impartiality arose around the desire for politically neutral reporting, yet even in political journalism, providing balance can be challenging when there are multiple parties or diverse views within parties. The situation is still more challenging for cross-cutting issues that transcend party divides. This is the case with many science-based controversies, where mainstream parties may all agree but a significant – or at least vocal – element of public opinion differs, or where the issue is too new for the mainstream parties even to have a position. Despite this, the exemplar of balance in a two-party political system still guides the reporting of such issues with the result that perspectives on a controversy are frequently reduced to two opposing sides. Furthermore, these opposing sides must be sufficiently different from each other to clearly represent differing opinions. Nuanced differences in opinion therefore tend to be neglected in favour of outright disagreement. Thus, a balanced-access approach derived from political reporting polarises and extremises complex issues, including issues drawing on scientific evidence.

Both objectivity and impartiality have been criticised as unattainable ideals that disguise biases within the news media: facts are inextricably laden with values; impartiality only operates within (sometimes quite narrow) cultural norms drawn from the wider society; and story selection inevitably entails implicit judgements about what is important. For instance, in an influential academic analysis of British television, the Glasgow Media Group argued that the ideal of objective and impartial news "merely leads to a 'naturalism' on the level of stylistics which hides its ideological components" (1980: p. 402). In the United States, in what have become known as the 'objectivity wars', some journalists have rejected the norm of objectivity altogether, charging it with passing off as neutral the biased standpoint of white privilege (Lowrey, 2020). Others, however, suggest that this rejection of objectivity mistakes process for outcome by attributing objectivity to the journalist rather than the method. As journalists Bill Kovach and Tom Rosenstiel put it in their *Elements of Journalism*, "objectivity has become a trap. The term has become so misunderstood and battered, it mostly gets the discussion off track" (2001: p. 41). Indeed, one of the traps is that the terms objectivity and impartiality – having evolved together from the shared impulse for independent, politically neutral journalism – are often used synonymously, yet, as Kovach and Rosenstiel again state clearly, "impartiality was never what was meant by objectivity" (p. 95).

On the one hand, objectivity describes an orientation towards reality that is independent of an individual's experience, an attempt to capture an external reality that persists regardless of what people think about it. Objectivity, then, is concerned with independently verifiable facts about the world. Accuracy, facticity, and, in a limited sense, truthfulness speak to this idea of objectivity. Impartiality, on the other hand, describes an orientation towards other people; in particular, how one treats the opinions and beliefs of others. Concepts such as neutrality, balance and non-partisanship share this orientation. So where objectivity is an epistemological principle concerned with verifiable knowledge, impartiality is a principle of justice; or, as BBC journalist Clive Myrie (2022) has said: "Impartiality is simply what's fair." In its orientation towards opinion, impartiality is concerned with the treatment of *subjective* realities. Objectivity and impartiality emerge from the same editorial impulse and both strive towards non-partisan truth-telling, yet in how they try to achieve this, they pull in exactly opposite directions.

To complicate matters further, as a reporting technique, balance can serve both objectivity and impartiality. While it can serve impartiality by acknowledging contrasting opinions, it serves objectivity by attributing factual claims to others in place of passing a judgement on their truth. Especially in the United States, where objectivity has been the primary journalistic norm, balance is often understood in this way as a "surrogate for validity checks" (Dunwoody and Peters, 1992: p. 210). Balance also has a third function: by portraying conflict, contrasting opinion enhances newsworthiness, especially if that opinion comes from a famous figure.

Of course, terms such as objectivity and impartiality are used loosely and flexibly in everyday discourse and their meanings have evolved historically. The point here is not to make a claim about correct usage; rather, it is to highlight that debates about journalistic standards rarely distinguish between two interconnected but distinct – and sometimes conflicting – ideals: objectivity is about telling the truth through a detached and unbiased gathering of the known facts; impartiality aims to tell the truth through a detached and unbiased gathering of perspectives. If objectivity aspires to the 'view from nowhere', as philosopher Thomas Nagel (1989) famously put it, and those who reject objective reporting aspire to a 'view from somewhere', as US journalist Lewis Raven Wallace has put it (2023), then impartiality aspires to 'views from everywhere'.

It is no coincidence that the journalistic norm of objectivity arose at the same time as another version of objectivity was formalised as the central component of the scientific method (Galison, 2015; Mindich, 1998; Schudson, 2001). Scientists, too, are interested in collecting verifiable facts about the world and developing theories that explain these facts. A scientist's personal route to a theory may be influenced by subjective factors, but in modern science, the aim is always that the final outcome is objective in that it captures an aspect of reality that is untainted by that personal journey (Daston and Galison, 2007). Significantly, however, scientists have no norm of impartiality in the sense of being inclusive of alternative perspectives. They are expected to give due credit to the work of others, but there is no expectation that they take account of subjective realities that might exist in addition to the objective facts. Indeed, science as a body of knowledge is partial in the sense that it admits no alternative to a scientific understanding of the world (even though individual scientists might simultaneously hold both scientific and religious views). This fundamental difference lies at the heart of the culture clash between scientists and journalists. As I discuss below in relation to the BBC's climate coverage, this has led to scientists implying that science is a special case where the norms of journalism should not apply.

Institutionalised impartiality at the BBC

Impartiality has always been a guiding principle for UK broadcasting. From its inception in 1922, the BBC was required to be both editorially independent of government and representative of the national culture in order for it to discharge its monopoly of access to the airwaves in a manner consistent with the public interest. Initially, the corporation was forbidden from covering political controversies, but it soon won the right to do so by promising, in the words of the then managing director John Reith, to "guarantee absolute impartiality" (Hanretty, 2009: p. 186). In practice, it was recognised that absolute impartiality was impossible and eventually it would be '*due* impartiality' that was written into the statutory requirements placed on the BBC. This required that the broadcaster include differing viewpoints when covering matters of public controversy.

Impartiality was seen as being closely tied to, but distinct from, objectivity. For example, the advisory paper of 1935 that first set out what impartiality meant at the BBC explained that it entailed the "objective presentation of all the alternative points of view" (Hanretty, 2009: p. 186). Over the following decades, the BBC's commitment to impartiality became a central part of its institutional identity. Public service broadcasters in many other countries followed the model making similar, albeit often more limited, commitments to impartial news reporting. Like US newspapers, the early BBC also instituted a clear separation between news and comment (see Chapter 2) and its current editorial guidelines on the impartial treatment of controversial subjects continue to require that: "Opinion should be clearly distinguished from fact" (BBC, 2023a).

Impartiality remains central to the BBC's core values and the BBC has frequently reflected on how it can be applied; a tendency that geneticist Steve Jones, in a review of the BBC's science coverage, describes as having a "theological intensity" (BBC Trust, 2011). The current Royal Charter that grants the BBC its license to operate lists as the first purpose of the corporation: "To provide impartial news and information" (BBC, 2016: p. 5). Perhaps surprisingly, this appeal to impartiality enfolds assumptions about cultural values, as the Charter goes on to direct that the corporation's news coverage is to be "firmly based on British values of accuracy, impartiality, and fairness" (p. 6). Indeed, impartiality has always been understood to be applicable only as long as it did not transgress the values that underpinned British democracy. Thus, speaking in the 1960s, the then Director General Sir Hugh Greene explained:

> In talking about the BBC's obligation to be impartial I ought to make it clear that we are not impartial about everything. There are, for instance, two very important exceptions. We are not impartial about crime ... nor are we impartial about race hatred.
>
> (Tulloch, 1990: p. 148)

Impartiality, then, was to work within the bounds of national consensus, even if vocal and perhaps sizable minorities disagreed with that consensus. These limits remain in place today, with the BBC guidance on impartiality stating explicitly that "the BBC's impartiality is not value free" and that "it incorporates the values of a democratic society", along with the unambiguous statement: "The BBC is not impartial on racism" (BBC, 2023b). This notion of impartiality being limited by a social consensus is particularly relevant when it comes to science reporting where, additionally, a scientific consensus may exist.

The corporation's current conception of impartiality has been influenced by a 2007 report commissioned by the BBC's then-governing body, the BBC Trust. According to the report author, independent programme maker John Bridcut, impartiality "involves a mixture of accuracy, balance, context, distance, evenhandedness, fairness, objectivity, open-mindedness, rigour, self-awareness, transparency and truth." To this already long list of essential constituents, Bridcut added "completeness" to emphasise the importance, as he saw it, of presenting a breadth of view (Bridcut, 2007, pp. 5-6). This did not mean providing equal space for every possible position, Bridcut explained, but "it should involve some space, provided that the points of view are rationally and honestly held, and all of them are subject to equal scrutiny" (p. 38).

For Bridcut, the breadth of views to be included, and the weight they should be given, was to be guided by public opinion. Thus, the impartial journalist needs to sense "where the preponderance of opinion lies on any issue [...]. It is the challenge for the content-provider to calibrate the range of views and their proportionate weight in the output" (Bridcut, 2007: p. 39). Bridcut noted that climate change was one subject where journalists needed to make these sorts of judgements, and where such judgements could be unpopular. He reported that a high-level seminar at the BBC about broadcasting on climate change had concluded that "the weight of evidence no longer justifies equal space being given to the opponents of the consensus." But, Bridcut added, "dissenters (or even sceptics) will still be heard, as they should, because it is not the BBC's role to close down this debate" (p. 40).

It is this desire for diversity that is representative of *public*, rather than *expert*, opinion that has led to concerns about some science coverage. Yet Bridcut's earlier inclusion of a proviso – "provided that the points of view are rationally and honestly held" – is significant. Public opinion was not the only factor to be considered. Notably, Bridcut's list of the essential elements of impartiality included accuracy, objectivity and truth. Yet it was *honesty*, not accuracy or facticity or veracity, that he identified as the main criterion by which to assess the legitimacy of an opinion – and one can honestly hold an opinion that is misguided, ill-informed or based on incorrect facts.

In addition to requiring due impartiality, the BBC also expects its output to display due accuracy, recognising this as a property distinct from due impartiality. Accuracy, the guidelines note, "is not simply a matter of getting facts right"; opinions may also need to be considered and all information weighed to reach the truth. Originally, the "due" caveat in both the impartiality and accuracy requirements indicated that the treatment would depend on the nature of the programme. Increasingly, however, it is "due weight" that is referenced. This resembles the "weight-of-evidence" strategy called for by science journalism scholar Sharon Dunwoody (2005), who has proposed that differing views should be reported but in a manner that shows where the bulk of the evidence and expert opinion lies. In the case of the BBC, the current guidelines for due impartiality state: "minority views should not necessarily be given similar prominence or weight to those with more support or to the prevailing consensus" (BBC, 2023a). However, the guidelines also go on to note that: "There may be occasions when the omission of views or other material could jeopardise impartiality. There is no view on any subject which must be excluded as a matter of principle, but we should make reasoned decisions, applying consistent editorial judgement, about whether to include or omit perspectives" (BBC, 2023a).

Contesting the BBC's climate coverage

Its sensitivity to the question of impartiality has made the BBC highly reflexive about its reporting of climate change. Between 1997 and 2004, senior BBC editors participated in a series of seminars with climate scientists and policy specialists. One participant recalls the BBC editors asking for the science and policy communities to be "more insistent and audible in the review of media performance" in order to shift the centre of gravity of the reporting (Smith, 2005: p. 1472). Similarly, an internal seminar, convened in 2006 as part of John Bridcut's re-evaluation of impartiality, included discussion of the extent to which impartiality should apply to climate reporting. Jana Bennet, then the BBC's Director of Television and formerly the head of science programming, advised her colleagues that, rather than "deciding some kind of fairness", they had a "duty to understand where the weight of the evidence has got to" (Bridcut, 2007: Appendix D: p. 103). Indeed, already in 2005, the Editorial Guidelines had specified, in the context of reporting risk, that: "If a contributor's view is contrary to majority scientific or professional opinion, the demands of accuracy may require us to make this clear" (BBC, 2005).

Despite this, in the years since, the BBC has frequently been accused of allowing impartiality to distort its coverage. This has especially been the case in its reporting of climate change where both institutional statements and specific broadcasts have provoked criticism. These include a 2011 report on the impartiality of the corporation's science coverage and news items featuring Nigel Lawson, the founding chairman of a climate sceptic lobby group and a former Conservative Chancellor of the Exchequer, whose newsworthiness as a famous figure ensured he was the preferred choice for contrarian comment.

The 2011 report was commissioned by the BBC's then-governing body, the BBC Trust, as one of a series of annual impartiality reviews of aspects of the BBC's output. The review consisted of a report by geneticist Steve Jones, a commentary by the BBC Trust and a response from the BBC Executive (BBC Trust, 2011). In his report, Jones drew on a series of interviews with BBC journalists and programme makers, scientists, politicians and others, and also on a content analysis of the BBC's science output commissioned by the BBC Trust which I led (Mellor et al., 2011). The content analysis had found little evidence of the use of balance across the BBC's news coverage of science as a whole, but balance was used more frequently in news stories dealing with controversial policy issues. About half of these broadcast reports, though only a quarter of such stories online, included a contributor making a critical comment about the science, whether pointing up a limitation of the science or making a more oppositional challenge. Those making oppositional comments were most likely to be non-scientists.

Alongside praising the BBC for the quality and breadth of its coverage, in his report Jones concluded that "false balance" had become widespread for some topics, with programmes "presenting the views of tiny and unqualified minorities as if they have the same weight as the scientific consensus" (BBC Trust, 2011: p. 55). Jones described hearing widespread concerns from the scientific community about the BBC's "insistence on bringing in dissident voices into what are in effect settled debates" (p. 55). Indeed, even before the review had got underway, commentators had already flagged up false balance as the key problem (Colquhoun, 2010; Henderson, 2010; Nature, 2010).

It was this charge of false balance that dominated responses to the review, with news headlines stating that the BBC had been found to give too much time to fringe views (e.g., Connor, 2011; Foster, 2011; Henderson, 2011; Revoir, 2011; Sample, 2011). Opinion pieces took a similar angle (e.g., Nutt, 2011; Toynbee, 2011). False balance continued to frame discussion of the BBC's coverage of climate change for years to come, in the United Kingdom and around the world; in 2014, for instance, an opinion piece in the *Washington Post* commented on a BBC update on the earlier impartiality review under the headline: "The distorting reality of 'false balance' in the media" (vanden Heuvel, 2014).

By contrast, responses to the review in the right-wing press claimed that the BBC was set to silence dissenting views. In a move that was itself an attempt at silencing, Nigel Lawson (2011) wrote that he had instructed his lawyers to issue the BBC with a letter demanding an apology for what he alleged was a false accusation against him in Jones' report. Lawson claimed that the BBC already had a clear policy not to include climate sceptics in reports about climate change.

In fact, the BBC Executive's response to the review had been more ambivalent than the accusations of censorship implied. It acknowledged that there would be times when non-scientific opinion was not relevant, but also other times "when it is appropriate to broadcast a range of views, including some from non-experts, because science cannot be divorced from the social, political and cultural environment in which it operates" (BBC Trust, 2011, pp. 9-10). The corporation promised to hold two seminars for staff on scientific issues and to establish an online training module on reporting science, which in the event ran for only about a year (Hope, 2017).

This action was not sufficient to stem criticisms and formal complaints. In 2014, a report by the parliamentary committee that oversees issues relating to science excoriated the BBC for "mistakes in their coverage of climate science by giving opinions and scientific fact the same weight" (House of Commons, 2014: p. 20). The report found that individual editors lacked the ability to determine the level of expertise of programme contributors and recommended that the BBC develop additional editorial guidelines for the coverage of climate change to enable presenters to challenge statements that departed too far from the scientific facts.

Just a couple of months later, the BBC was found to have mishandled an interview on the flagship radio news programme *Today*, which had pitched Nigel Lawson against Brian Hoskins, a professor of meteorology and Director of the Grantham Institute for Climate Change in London. The BBC had initially defended itself by justifying Lawson's inclusion in the programme on the grounds that it was legitimate for him to debate the economic implications of climate policies (Pidcock, 2014). Complainants claimed that in fact Lawson spoke about the science; in particular, he had rejected Hoskin's summary of the evidence showing an increase in extreme weather events. The BBC's Editorial Complaints Unit found that the programme had not made it sufficiently clear that Lawson's views were not supported by the research, and that his treatment within the item had implied that his views were on an equal footing with Hoskins' views (ECU, 2014). The corporation's response was again to offer senior journalists the opportunity to attend (the now defunct) training course. Lawson responded by saying he'd been banned and again accused the "Stalinist" BBC of silencing debate about climate (Lawson, 2014).

A similar sequence of events played out in 2017 when the Editorial Complaints Unit upheld a complaint from the Green Party about another Lawson interview on the *Today* programme. The programme makers had defended the interview on the grounds of providing impartiality: "The BBC's role is to hear different views so listeners are informed about all sides of debate and we are required to ensure controversial subjects are treated with due impartiality" (BBC, 2017). Again, the corporation justified Lawson's inclusion due to his economic expertise. On this occasion, the complaint also went to the broadcast regulator, Ofcom, which found that the interview had breached the Broadcast Code (Ofcom, 2018). Ofcom found that the programme's failure to challenge scientific claims made by Lawson in the course of the interview had led to significant inaccuracies and had therefore breached the requirement for due accuracy.

Likely in response to the Ofcom ruling, the following year the BBC introduced another training course, this time specifically on reporting climate change. A leaked email about the new course from the BBC's director of news and current affairs, Fran Unsworth, included a "crib sheet" on climate reporting. This began by admitting that the BBC got its climate coverage "wrong too often". Second on a list itemising the BBC's position, after a statement accepting that anthropogenic climate change is happening, was the advice that journalists should: "Be aware of 'false balance'" (Hickman, 2018). Casting the Intergovernmental Panel on Climate Change in the role of referee, the memo continued:

> To achieve impartiality, you do not need to include outright deniers of climate change in BBC coverage, in the same way you would not have someone denying that Manchester United won 2-0 last Saturday. The referee has spoken. However, the BBC does not exclude any shade of opinion from its output, and with appropriate challenge from a knowledgeable interviewer, there may be occasions to hear from a denier.
>
> (BBC, cited in Hickman, 2018)

Once again, the BBC was seeking a formulation that simultaneously recognised the scientific consensus as the best statement of fact and a reliable source of objective knowledge, while still leaving room for the corporation to discharge its democratic function in giving voice to a wide range of public opinion.

As the 2017 Ofcom ruling illustrates, complaints that originate as concerns about false balance often transform into complaints about the presence of incorrect claims by the time they are upheld by regulators. As Bob Ward, Policy and Communications Director at the Grantham Research Institute for Climate Change, pointed out on the eve of the publication of the 2011 impartiality review: "the real challenge for the BBC is not just to ensure the impartiality of its coverage of climate change, but also to uphold its equally important commitment to accuracy" (Ward, 2011). In its attempts to be inclusive of a diversity of views around scientific issues, the BBC risks introducing factual inaccuracies. The otherwise similar norms of impartiality and objectivity seem to be in conflict. This is the core difficulty that journalists must navigate.

From false balance to scientific exceptionalism

Steve Jones' review of the impartiality of the BBC's science coverage recommended that the BBC take "a less rigid view of 'due impartiality' as it applies to science" (BBC Trust, 2011: p. 17). Some of the people he had interviewed in the course of compiling his report – specifically, some of those with science backgrounds – had proposed that the BBC guidelines should be amended to exempt science from some of its editorial policies (p. 76). Jones himself thought this went too far as it could provoke allegations of censorship, like those since witnessed around climate reporting. Yet Jones, too, expected science to be treated as a special case, albeit through "the application of common sense" rather than through formal editorial directives. To Jones and other scientists similarly frustrated by news outlets' use of balance, science is unique in that the processes of organised science already ensure that the output of science is objective and evidenced. As Jones put it, where many fields tolerate – or even welcome – opposing opinions, "science deals, most of the time, with opposed *evidence*" (p. 56; emphasis added). Journalists, he implies, should trust what scientists say.

Yet evidence only becomes opposed to other evidence if it has been subjected to someone's opinion – albeit an expert opinion – in an act of interpretation that judges the applicability, relevance and robustness of each finding (see, e.g., Hilgartner, 2000). Jones acknowledged this by allowing that it could be relevant to use a balance of voices in science journalism. Like those who frame their concerns in terms of 'false equivalence', he proposed that balancing comments could be appropriate provided they only came from experts. He called this exclusionary approach an "equality of voice": "a match of scientists not with politicians or activists, but with those qualified to take a knowledgeable, albeit perhaps divergent, view of research" (p. 16). To scientists such as Jones, then, impartiality and objectivity are in conflict because science is already objective. The conflict is resolved, the suggestion is, by accepting that only scientists can challenge scientists.

The difficulty is that the processes of peer review are not infallible and that, in any case, scientists do not only comment on scientific knowledge that has passed through such tests. Scientists also explain how the science is relevant to policy decisions; they consider emerging science that has not yet had time to resolve into a consensus (as was most evident in the early stages of the Covid pandemic); and, not infrequently, they claim that the scientific evidence supports a particular policy direction. These are not matters that are reducible to objective statements that are provable through experimental or mathematical methods; they involve value judgements and opinion. To exempt such claims from challenge by politicians or campaigners, who may have a much better understanding of the social consequences or moral limits of policy positions advocated by scientists, would be a significant restriction on democratic debate. An approach to balance that was based on scientific exceptionalism would privilege narrow technical expertise over broader contextual understandings.

Conclusion

Journalists use balance for multiple reasons: to construct newsworthiness and to satisfy the journalistic norms of both objectivity and impartiality. In serving these two separate norms, balance reinforces the confusion between the epistemological principle of objectivity, with its focus on accuracy and facticity, and the democratic principle of impartiality, with its focus on fairness and diversity of media access. The concept of 'false balance', in using the language of falsity, foregrounds the use of balance as a substitute for, and therefore a threat to, objectivity and simultaneously obscures the use of balance to ensure fair representation.

The BBC, like other news outlets, has tried to walk a line between its twin obligations of accuracy and impartiality. Its editorial policy acknowledges that balance is just one mechanism through which impartiality may be achieved, that impartiality has limits and that accuracy is also essential (BBC, 2023a). Yet, for over a decade, scientists and climate campaigners have complained that the corporation's interpretation of its impartiality requirements has led to the use of false balance in its climate coverage by opposing the accurate statements of climate scientists with the inaccurate claims of climate sceptics. Climate sceptics, on the other hand, have complained that the corporation is failing to be impartial by censoring contrarian voices.

Journalists at outlets around the world, including at the BBC, have often applied balance in ways that have allowed misleading claims about climate change to get airtime. Yet to characterise this as false balance is to imply that journalists overuse balance. One of the ironies of the commentary about impartiality is that, in fact, balance is rarely used in science journalism. For instance, the content analysis supporting the BBC's science review found that only one third of broadcast science news reports included comments from more than one contributor; even when more than one contributor was included, this often came from sources connected to the research or the research organisation (Mellor et al., 2011). Most of this science news was sourced from press releases, meaning that the BBC's science news rarely balances a source's promotional perspective with an alternative, disinterested perspective. Evidence for churnalism at other outlets (see chapter 4) suggests that the BBC might not be unusual in avoiding balance when reporting new research findings.

Another irony is that scientists, whom one might expect to value impartial analysis, have often taken up positions that seem to reject impartial journalism even as an ideal. In settling on the convenient slogan of 'false balance', the critics of journalistic impartiality invoke a logic that appeals to scientific exceptionalism. Impartiality is not the only aspect of media standards where it has been proposed that science should be treated as a special case: during a revision of UK libel law, scientists and others petitioned – with partial success – for scientific claims to be exempted from possible libel actions (Mullis and Scott, 2014). In the case of impartiality, exempting scientific news sources could place further limits on journalistic scrutiny in a field where there is already relatively little questioning of news sources. By construing the problem as a failure to respect the special status of scientific knowledge, the concept of false balance misdirects attention away from the manipulative practices of lobbyists (including some scientists) acting on behalf of powerful vested interests, towards journalists' supposed inability to understand scientific method.

The BBC is highly exposed to political pressure and commercial competition, leaving it little room to manoeuvre between losing its independence and losing its audience. A public service broadcaster like the BBC does not have the luxury to ignore public opinion. When social media channels amplify organised climate denialism – or other campaigns that reject a scientific consensus – to the extent that a significant proportion of the population is exposed to such messages, the BBC has to address this in some way. Balanced, objective and impartial reporting need not mean allowing inaccurate claims to pass unchallenged, nor allowing unacknowledged vested interests to be promoted. Indeed, those traditional journalistic norms can be used to expose the manipulation of public opinion, as some broadcasters – including the BBC – have recognised in appointing specialist disinformation correspondents.

Balance can be applied thoughtfully – for instance, to draw out the assumptions behind a scientific claim, or to expose the manipulation of public opinion. Or it can be applied poorly – as an unthinking recourse to over-exposed commentators who polarise public debate. The lack of scrutiny of scientific claims in routine science reporting leaves science journalism ill-prepared to deal with controversial policy issues where science is an essential ingredient in good policymaking, but where it can also become a resource to be continually un-done in order to delay policy action. The question is not whether balance is a valid journalistic method in the context of science, as those who complain about false balance imply, but whether the lack of balance in routine science news and the perfunctory use of balance in reporting science-based policy issues make science journalism unusually vulnerable to becoming a platform for the strategic communications of self-interested groups.

References

Allan, S., 1997. News and the public sphere: Towards a history of objectivity and impartiality. In Bromley M. and O'Malley, Tom. (eds.), *A Journalism Reader*, London: Routledge, pp. 296–329.

BBC, 2005. *Editorial Guidelines: The BBC's Values and Standards*. London: BBC.

BBC, 2016. *Royal Charter for the Continuance of the British Broadcasting Corporation*. Cm 9365, https://www.bbc.com/aboutthebbc/governance/charter.

BBC, 2017. BBC defends Lord Lawson climate change interview. *BBC News*, 10 Aug., https://www.bbc.co.uk/news/uk-politics-40889563.

BBC, 2023a. Guidelines. *Editorial Guidelines*. BBC, https://www.bbc.com/editorialguidelines/guidelines.

BBC, 2023b. Guidance. *Editorial Guidelines*. BBC, https://www.bbc.co.uk/editorialguidelines/guidance.

BBC Trust, 2011. *Review of Impartiality and Accuracy of the BBC's Coverage of Science*. London: BBC Trust, https://www.bbc.co.uk/bbctrust/our_work/editorial_standards/impartiality/science_impartiality.html.

Boykoff, M.T., 2007. Flogging a dead norm? Newspaper coverage of anthropogenic climate change in the United States and United Kingdom from 2003 to 2006. *Area*, 39(4), pp. 470–481.

Boykoff, M.T., 2008. Lost in translation? United States Television news coverage of anthropogenic climate change, 1995–2004. *Climatic Change*, 86(1), pp. 1–11.

Boykoff, M.T. and Boykoff, J.M., 2004. Balance as bias: Global warming and the US prestige press. *Global Environmental Change*, 14(2), pp. 125–136.

Boykoff, M., and Farrell, J., 2019. Climate change countermovement organizations and media attention in the United States. In: Almiron, N. and Xifra, J. (eds.), *Climate Change Denial and Public Relations: Strategic Communication and Interest Groups in Climate Inaction*. London: Routledge, pp. 121–139.

Boykoff, M.T. and Mansfield, M., 2008. 'Ye olde hot Aire': Reporting on human contributions to climate change in the UK tabloid press. *Environmental Research Letters*, 3, 10.1088/1748-9326/3/2/024002.

Bridcut, J., 2007. *From Seesaw to Wagon Wheel: Safeguarding Impartiality in the 21st Century*. London: BBC Trust.

Collins, H., 1987. Certainty and the public understanding of Science: Science on television. *Social Studies of Science*, 17(4), pp. 689–713.

Colquhoun, D., 2010. Science reporting on the BBC. Your chance to have a say. *DC's Improbable Science*. 19 Sep., http://www.dcscience.net/2010/09/19/science-reporting-on-the-bbc-your-chance-to-have-a-say.

Connor, S., 2011. Climate sceptics get too much air-time, BBC told. *The Independent*, 21 Jul, p. 16.

Cushion, S., 2012. *The Democratic Value of News: Why Public Service Media Matter*. Basingstoke: Palgrave Macmillan.

Daston, L. and Galison, P., 2007. *Objectivity*. Brooklyn: Zone Books.

Dayrell, C. and Urry, J., 2015. Mediating climate politics: The surprising case of Brazil. *European Journal of Social Theory*, 18(3), pp. 257–273.

Dunwoody, S., 2005. Weight-of-evidence reporting: what is it? Why use it? *Nieman Reports*, 15 Dec., https://niemanreports.org/articles/weight-of-evidence-reporting-what-is-it-why-use-it/.

Dunwoody, S. and Peters, H.P., 1992. Mass media coverage of technological and environmental risks: A survey of research in the United States and Germany. *Public Understanding of Science*, 1(2), pp. 199–230.

ECU, 2014. *Governor's Bulletin: Disaggregated Data*. BBC Executive Complaints Unit, http://downloads.bbc.co.uk/mediacentre/ecu_apr14_sep14.pdf.

EJN, 2022. Our five core principles. Ethical Journalism Network, https://ethicaljournalismnetwork.org/who-we-are.

Foster, P., 2011. Climate change sceptics to get less coverage, rules BBC. *The Daily Telegraph*, 19 Jul., p. 3.

Galison, P., 2015. The journalist, the scientist and objectivity. In: Padovani, F., Richardson, A. and Tsou, J.Y. (eds), *Objectivity in Science: New Perspectives from Science and Technology Studies*. Cham: Springer.

Glasgow Media Group, 1980. *More Bad News*. London: Routledge and Kegan Paul.

Hanitzsch, T. et al., 2011. Mapping journalism cultures across nations: A comparative study of 18 countries. *Journalism Studies*, 12(3), pp. 273–293.

Hanretty, C., 2009. *The Political Independence of Public Service Broadcasters*. PhD thesis. European University Institute.

Henderson, M., 2010. Dear BBC: balance isn't everything. *The Times*, 4 Feb., p. 13.

Henderson, M., 2011. BBC criticised for 'bogus impartiality' favouring dissident science views. *The Times*, 20 Jul., https://www.thetimes.co.uk/article/bbc-criticised-for-bogus-impartiality-favouring-dissident-science-views-9lph78mbncn.

Hickman, L., 2018. BBC issues internal guidance on how to report climate change. *CarbonBrief*, 7 Sep., https://www.carbonbrief.org/exclusive-bbc-issues-internal-guidance-on-how-to-report-climate-change/.

Hilgartner, S., 2000. *Science on Stage: Expert Evidence as Public Drama*. Stanford: Stanford University Press.

Hope, M., 2017. No BBC science reporting course since 2012, doc reveals. *DeSmog*, 3 Nov., https://www.desmog.com/2017/11/03/no-bbc-science-reporting-course-2012-docs-reveal.

House of Commons, 2014. *Communicating Climate Science*. Eighth Report of Session 2013–14, HC 254. London: The Stationery Office.

Kovach, B. and Rosenstiel, T., 2001. *The Elements of Journalism: What News-People Should Know and the Public Should Expect*. New York: Three Rivers Press.

Latour, B. and Woolgar, S., 1986. *Laboratory Life: The Construction of Scientific Facts*. Princeton: Princeton University Press.

Lawson, N. 2011. Listen up, BBC, the climate debate is far from over; A BBC Trust report said only the 'deluded' did not accept harmful global warming. Nigel Lawson and numerous experts beg to dissent. *The Times*, 16 Oct., p. 5.

Lawson, N. 2014. 'I've been banned by the BBC!': Ex-Chancellor Lord Lawson, a passionate climate change sceptic, accuses BBC bosses of silencing debate on global warming. *Mail Online*, 9 Jul., https://www.dailymail.co.uk/debate/article-2685405/Ive-banned-BBC-Ex-Chancellor-Lord-Lawson-passionate-climate-change-sceptic-accuses-BBC-bosses-silencing-debate-global-warming.html.

Lowrey, W., 2020. A reckoning over objectivity, led by Black journalists. *New York Times*, 23 Jun., https://www.nytimes.com/2020/06/23/opinion/objectivity-black-journalists-coronavirus.html.

McAllister, L., Daly, M., Chandler, P., McNatt, M., Benham, A. and Boykoff, M., 2021. Balance as bias, resolute on the retreat? Updates & analyses of newspaper coverage in the United States, United Kingdom, New Zealand, Australia and Canada over the past 15 years. *Environmental Research Letters*, 16, doi.org/10.1088/1748-9326/ac14eb.

Mellor, F., Webster, S. and Bell, A.R., 2011. *Content Analysis of the BBC's Science Coverage*. London: BBC Trust, http://downloads.bbc.co.uk/bbctrust/assets/files/pdf/our_work/science_impartiality/appendix_a.pdf.

Michaels, D., 2008. *Doubt Is Their Product: How Industry's Assault on Science Threatens Your Health*. Oxford: Oxford University Press.

Mindich, D.T.Z., 1998. *Just the Facts: How 'Objectivity' Came to Define American Journalism*. New York: New York University.

Mooney, C., 2010. Blinded by science: How 'balanced' coverage lets the scientific fringe hijack reality. *Columbia Journalism Review*, 43(4), https://www.discovermagazine.com/the-sciences/blinded-by-science-how-balanced-coverage-lets-the-scientific-fringe-hijack-reality.

Mullis, A. and Scott, A., 2014. Tilting at windmills: The defamation act 2013. *The Modern Law Review*, 77(1), pp. 87–109.

Myrie, C., 2022. Opinion is cheap and easy. The BBC's mission is to deliver facts and evidence. *The Guardian*, 4 Sep., https://www.theguardian.com/commentisfree/2022/sep/04/opinion-is-cheap-and-easy-bbc-mission-is-to-deliver-facts-and-evidence.

Nagel, T., 1989. *The View from Nowhere*. Oxford: Oxford University Press.
Nature, 2010. The balance of probabilities. Editorial. *Nature*, 467, 21 Oct., p. 883.
Nelkin, D., 1987. *Selling Science: How the Press Covers Science and Technology*. New York: W.H. Freeman.
Nutt, D. 2011. Review was right to criticise BBC's science coverage. *The Guardian*, 22 Jul., https://www.theguardian.com/commentisfree/2011/jul/21/review-bbc-science-coverage.
Ochs, A.S., 1896. Business announcement. *New York Times*, 18 Aug., https://timesmachine.nytimes.com/timesmachine/1896/08/19/103384582.pdf.
Ofcom, 2018. *Ofcom Broadcast and On Demand Bulletin*. Ofcom, 351, 9 Apr., https://www.ofcom.org.uk/__data/assets/pdf_file/0012/112701/issue-351-broadcast-on-demand-bulletin.pdf.
Ofcom, 2021. *The Ofcom Broadcasting Code*, https://www.ofcom.org.uk/tv-radio-and-on-demand/broadcast-codes/broadcast-code.
Oreskes, N. and Conway, E.M., 2010. *Doubt Is Their Product: How A Handful of Scientists Obscured the Truth on Issues from Tobacco Smoke to Global Warming*. New York: Bloomsbury Press.
Painter, J., 2011. *Poles Apart: The International Reporting of Climate Scepticism*. Oxford: Reuters Institute for the Study of Journalism.
Pidcock, 2014. BBC upholds complaint over Today Programme Nigel Lawson interview. *Carbon Brief*, 7 Jul., https://www.carbonbrief.org/bbc-upholds-complaint-over-today-programme-nigel-lawson-interview.
Revoir, P., 2011. BBC move to sideline climate sceptics. *Daily Mail*, 20 Jul., p. 17.
Sample, I., 2011. BBC giving too much weight to fringe views in science, review finds: 'False balance' in coverage of climate, GM and MMR. *The Guardian*, 21 Jul., p.17.
Scannell, P. and Cardiff, D., 1991. *A Social History of British Broadcasting*. Volume 1. Oxford: Blackwell.
Schudson, M., 1981. *Discovering the News: A Social History of American Newspapers*. New York: Basic Books.
Schudson, M., 2001. The objectivity norm in American journalism. *Journalism*, 2(2), pp. 149–170.
Smith, H.K., 1965. The deadly balance. *Columbia Journalism Review*, 4(3), p. 13.
Smith, J., 2005. Dangerous news: Media decision making about climate change risk. *Risk Analysis*, 25(6), pp. 1471–1482.
Toynbee, P., 2011. As unreason grips America, beware our flat-earthers too. *The Guardian*, 2 Aug., p. 27.
Tulloch, J., 1990. Television and black Britons. In Goodwin, A. and Whannel, G. (eds.), *Understanding Television*. London: Routledge.
vanden Heuvel, K., 2014. The distorting reality of 'false balance' in the media. *Washington Post*, 15 Jul., p. 14.
Wallace, L.R., 2023. *The View from Somewhere: Undoing the Myth of Journalistic Objectivity*. Chicago: University of Chicago Press.
Ward, B., 2011. BBC coverage of climate change should be accurate as well as impartial. Grantham Institute, 19 Jul., https://www.lse.ac.uk/granthaminstitute/news/bbc-coverage-of-climate-change-should-be-accurate-as-well-as-impartial.

9

MAKING SCIENCE TREND

The style of popular science magazines

Katarzyna Molek-Kozakowska

The 2021 satirical film *Don't Look Up* lampooned the consequences of a failure to communicate scientific findings to political leaders, media influencers and the public at large. Conceived as an allegory for the impending climate catastrophe, the film posits that celebrity culture and its attendant media logic, driven by rapid attention shifts and the market for entertaining stories, serve to dismiss important science even at a time of existential threat. The film tells the story of two astronomers, untrained in dealing with the media, who turn to news outlets in order to warn that a comet they have discovered will soon impact Earth and destroy all life on the planet. The media respond by framing the discovery as an odd curiosity. Rather than explaining the risk that the comet poses, as the scientists had hoped, the focus shifts to unrelated issues, such as the scientists' personal lives; when the risk is reported, the uncertainty of the disaster happening is exaggerated to match a populist political agenda.

As journalism scholar Declan Fahy has argued, the film's caricature of a morally corrupted and trivialising news media disregards the steady work of journalists who have reported on climate change for decades (Fahy, 2022). Yet its heavy-handed satire notwithstanding, some science communicators lauded the film for its accurate depiction of the central challenge for science journalism: the mismatch between the communication strategies of scientists and journalists (Amsen, 2022; Angelone, 2022; Steenstrup-Duch, 2022). On the one hand, scientists, accustomed to communicating with their peers, are inclined to focus on content. They emphasise what they know and how they know it, as do the astronomers in the film when they try, unsuccessfully, to persuade the US President of the danger the world faces by explaining the method that led them to identify the comet. Journalists, on the other

DOI: 10.4324/9781003326724-13

hand, prioritise catching the audience's attention and generating emotional responses through well-honed journalistic techniques.

The challenges of "accommodating science" (Fahnestock, 1993) to the needs and preferences of non-expert audiences have been studied by language and rhetoric scholars. Recommendations to simplify the science (but not too much), to reduce jargon (without eliminating key terms), to provide additional explanations and reminders of the basic science (while still emphasising what is new) and to sharpen the conclusions and implications to boost relevance (but not to overgeneralise and distort them) are often repeated in blogs and other publications (Olson, 2015). Such suggestions involve treading a fine line between style and substance.

Style matters because it is not at all easy to present scientific content in attractive ways that foster *both* understanding *and* engagement. Audience-attracting stylistic techniques can inhibit the full communication of scientific evidence and reasoning. Yet without these techniques, it is difficult to engage the audience – and without engagement, communication fails. The ability to strike the right balance is further aggravated by the potential for some strategic devices, used by both commercial news outlets and science PR, to have a rhetorical effect beyond simply attracting an audience. In pursuit of eyes and clicks, stylistic choices have the potential, in the long term, to diminish authority and trust in science by misrepresenting complex and nuanced information or by hyping up extraordinariness, while de-emphasising essential information to the point that it goes unreported.

This chapter considers the linguistic styles, rhetorical devices and storylines through which the journalistic voice is constructed. I identify three distinct stylistic shifts that journalists accomplish as they adapt science from academic discourse to the popular science genre. I call these the 'attentional turn', the 'affective turn' and the 'narrative turn'. It is important to stress that, while it is useful for analytical purposes to separate out these three strategies, journalists do not use one or another. Rather, they deploy these stylistic resources in a combined and skilful manner in order to draw more readers to their reports.

I illustrate the three stylistic shifts with a set of examples drawn from *New Scientist*, the UK-based magazine that describes itself as "the world's most popular weekly science and technology publication" (New Scientist, 2023). The 25 articles in this sample all appeared on the magazine website as "trending" headlines in the summer of 2022; as such, they represent articles that attracted high levels of audience engagement.[1] This sample complements a larger sample of headlines from "most-read" articles in the same magazine that I analysed in an earlier study and that showed how *New Scientist* uses conventional journalistic devices to attract audiences (Molek-Kozakowska, 2016).

The trending articles drawn on here display a range of stylistic patterns and report on a variety of topics. Many cover new discoveries and research results, particularly relating to medical advances or new space images, but health and well-being, environmental research on climate and biodiversity and artificial intelligence (AI) technologies are also covered. Thematically, this reader-driven selection is similar to preferences for science journalism content more generally (e.g., Molek-Kozakowska, 2017; Weitkamp, 2003) and aligns with the selection criteria applied by editors at other popular science publications (Guenther and Ruhrmann, 2013).

The attentional turn: Newsworthiness

A long tradition of media scholarship has examined what sort of events get reported due to their intrinsic newsworthiness. Starting with a foundational 1965 study of foreign news by Norwegian media researchers Johan Galtung and Mari Ruge, scholars have listed the 'news values' (or 'news factors' in Galtung and Ruge's terminology) that guide journalists' story selection; criteria that are in turn informed by journalists' and editors' perceptions of what types of stories appeal to their readers or viewers. Events that would be considered inherently newsworthy in general news coverage might include, for example, natural disasters, political accords or heroic deeds.

However, news values do not reside only within the events being reported but also in the manner in which journalists report on events. Indeed, events that may be low in inherent newsworthiness may be made newsworthy through a selection of stylistic and rhetorical devices (Bednarek and Caple, 2014). These stylistic expressions of news values are often made most strongly early in the news text – in the headline and in the lead paragraph, which typically unpacks the vague or dense headline (Bednarek and Caple, 2017; Bell, 1991). The more news values are brought into the headline and the lead, the higher the likelihood of catching the attention of a reader. This is not to claim that style alone can make someone pay attention to a topic they have no time for, but it can help spur their interest or curiosity.

The common technique of 'clickbait' in online journalism exemplifies this practice. Clickbait headlines capitalise on novelty, ambiguity or forward-reference in order to lure readers into a story, even if the story itself is trivial or already well-known (Blom and Hansen, 2015). Through stylistic features such as unusual punctuation or hyperbolic language, clickbait aims to capture reader attention and increase virality. For instance, one common technique is to pose a question in a headline; almost a tenth of headlines in a sample of 'most-read' articles in the UK's *Daily Mail* tabloid used this device (Molek-Kozakowska, 2013). Other studies have found that posing questions in headlines decreases, rather than increases, reader interaction and the click-through rate among social media consumers of news (Jung et al., 2022). This may be explained by a reluctance to share information that is perceived as having lower credibility, as is signalled when the topic has been introduced via a question. As this suggests, the stylistic repertoire used to enhance newsworthiness reflects, to varying degrees, journalistic assumptions about audience preferences as well as actual reader preferences.

Studying the form or composition, rather than just the content, of journalistic texts can reveal how newsworthiness is projected onto events that would not otherwise draw public attention, as is the case for many developments in science. This entails examining not the "what" of popular news – such as politics or celebrity, scandal or breakthrough – but rather the "how" of the stylistic resources responsible for making a piece relevant, attractive and worth investing time and energy in reading, whatever the content preferences of the audience. Such an approach to newsworthiness allows us to see news values as a combination of specific references to sources and actors and of engaging lexical choices and formulaic expressions (Bednarek and Caple, 2017: p. 55; see Molek-Kozakowska and Struchkova, 2022, for science-oriented news). In addition to lexical choices of which words to use, journalists can also deploy a variety of rhetorical resources that construct newsworthiness indirectly; for instance, through wordplay, allusion, comparison, metaphor or irony. Visuals are also used to attract attention and enhance newsworthiness (Bednarek and Caple, 2014).

Scholars have examined general news coverage in order to identify the different news values at play. The resulting typologies often differ in how they name the various news values, but they are broadly similar in the values they identify (Bednarek, 2016). In the following typology, which I will use to classify the examples from *New Scientist*, I indicate how each of the news values can be constructed through journalists' lexical and grammatical choices. Similar stylistic resources could be identified for other typologies.

Consonance includes reference to expected outcomes and recurrent situations or stereotypes, sometimes signalled with 'again' or 'as predicted'.

Eliteness is a reference to well-known individuals, institutions, countries or events.

Impact is realised with verbs of action or change ('transform'), evaluations of significance ('major'), commitment ('must have') or conditionality (if/then).

Negativity foregrounds undesirable and unpleasant situations, outcomes or feelings expressed through pejorative words or phrases, as well as the grammatical negation of positives.

Personalisation involves individuals' perspectives and stories, as in the focus on the person rather than the issue, or a direct address to the reader ('you').

Positivity relies on listing advantages, desirable outcomes, values and positive developments.

Proximity mentions locations, facts and values that are geographically close or culturally relevant.

Superlativeness is expressed by modifiers intensifying particular aspects of an event ('extreme', 'never before') and through the use of large numerals or quantifiers ('masses') and superlatives ('the most significant').

Timeliness involves indications of recency ('today') and is also indicated through verb tense (present) and aspect (continuous), or through references to an ongoing process.

Unexpectedness expresses not only surprise ('unexpectedly') but also newness, and is often lexicalised in terms of deviance and rarity.

In its need to attract an audience, science journalism is no different from other forms of journalism that draw on these news values. A key task for science journalists, then, is to render science newsworthy. In an empirical study of four German daily newspapers, Franziska Badenschier and Holger Wormer (2012) concluded that, with some modifications and additions, the same news values identified in general (political) news, such as the list above, can also be found in science coverage. However, the infrequency of must-know events or extraordinary breakthroughs in science means that constructing newsworthiness through stylistic means is particularly important for science journalism.

The headlines of *New Scientist*'s trending articles show how highly read articles emphasise news values through specific word choices. In Table 9.1, the underlined words and phrases encode a news value, identified in the second column that the headline projects onto the story.

For example, the headline "How *Minecraft* is helping children with autism make new friends" stylistically exploits an incongruity between what we know about "being autistic" and "making friends". It also points us to an unexpected technological ally – a game – to facilitate remedial or therapeutic action. The interest and relevance of the piece are further increased through Superlativeness and Unexpectedness in the third paragraph:

> So, in 2013, Duncan, a web developer in Timmins, Canada, set up a server to run a version of *Minecraft* exclusively for children with autism and their families. He thought the invite-only server would attract 10 or 20 people. To his surprise, hundreds requested to join in the first few days.
>
> (Rutkin, 2016)

Mass unspecified numbers ("hundreds") mentioned in the context of "surprise" imply not only the existence of a significant unassessed need for such games but also the potentially high positive impact on the community. This is mixed with affective and narratival appeals through a mini-story of the developer with a touch of authentic detail signalled by the precise chronology. As discussed later, attributing thoughts and emotions to characters in a story enhances the newsworthiness by creating a relatable and compelling read.

TABLE 9.1 Sample of *New Scientist* headlines of articles trending in the summer of 2022, with coding for news values; underlined phrases indicate how the news value is expressed in the headline.

Article headline	*News value in headline*
How Minecraft is helping children with autism make new friends	Unexpectedness
Is our solar system a cosmic oddity? Evidence from exoplanets says yes	Consonance
Grape seed chemical allows mice to live longer by killing aged cells	Impact
We are finally starting to understand brain fog and how to treat it	Timeliness
Long Covid symptoms may include hair loss and ejaculation difficulties	Negativity
DeepMind's protein-folding AI cracks biology's biggest problem	Superlativeness
Your essential guide to the many breath-taking wonders of the universe	Positivity
40°C heatwave may have killed 1000 people in England and Wales	Negativity
Ultrasound stickers could continuously image internal organs for days	Timeliness
Small woodlouse-like crustacean 'pollinates' red seaweed with its legs	Unexpectedness
Blood substitute repairs damaged organs hours after heart stops	Impact
Cold exposure stops tumour growth in mice by hijacking glucose stores	Impact
How the secrets of ancient cuneiform texts are being revealed by AI	Timeliness
JWST has caught two galaxies smashing together and sparking starbursts	Superlativeness
Transfusing blood from an old mouse to a younger mouse causes ageing	Consonance
Hummingbird that was feared extinct is spotted in Colombian mountains	Unexpectedness
If cryptocurrencies are unhackable, how do they keep getting stolen?	Unexpectedness
How to spot the 2022 Perseid meteor shower that will peak on Saturday	Personalisation
What Earth's mysterious infancy tells us about the origins of life	Proximity
The power of quiet: The mental and physical health benefits of silence	Positivity
Ignition confirmed in a nuclear fusion experiment for the first time	Timeliness
Cornea made from pig collagen gives people who were blind 20/20 vision	Impact
Why thinking hard for several hours can leave you mentally exhausted	Personalisation
Simple chemistry can recycle polystyrene into more valuable products	Impact
Faint distant galaxies spotted by JWST may be closer than they appear	Proximity

A commonly practised way to make a piece of science news highly attractive is to present it as revolutionary. For example, an article on protein folding makes the most of the publication of a database, which was announced by the AI-software company DeepMind and introduced as UniProt, the Universal Protein Knowledgebase (UniProt Consortium, 2022). The completion of a costly project and the accompanying data release are presented as extraordinary through many language choices:

> Artificial intelligence firm DeepMind has transformed biology by predicting the structure of nearly all proteins known to science in just 18 months, a breakthrough that will speed drug development and revolutionise basic science.
>
> [...]
>
> Today, the company announced that it is publishing the structures of more than 200 million proteins – nearly all of those catalogued on the globally recognised repository of protein research, UniProt.
>
> (Sparkes, 2022)

To stress the significance of the scientific step taken, editors use verbs of Impact ("transform", "revolutionise") and charged words that connote a discovery ("breakthrough"). In addition, with Timeliness ("today") and Superlativeness explained in relatable figures ("just 18 months", "200 million"), the global significance of the publication is asserted. Through such language, the article encourages readers to appreciate the immediate and tangible contribution of this company to future medical research.

As this example also suggests, news values are not constructed by journalists alone. Science journalism is highly reliant on information subsidies such as press releases and press conferences (see Chapter 4). These often deploy the same news values in order to provide journalists with easy-to-reuse material while also promoting the source's activities. In the case of the DeepMind story, the company used similar lexical choices as did the news coverage – "breakthrough", "world's leading" – to announce their database (UniProt Consortium, 2022). The stylistic enhancement of newsworthiness can therefore result in news coverage that uncritically accepts a framing of developments that forwards the source's interests. Indeed, news values become weapons in what Marta Entradas and Martin Bauer (2022) have referred to as "an arms race for visibility" within science.

According to German sociologist Peter Weingart (2022), this constitutes a marked shift from "truth value" to "news value" that must be acknowledged and confronted. The attention-seeking orientation of journalism has become embedded in science itself, in what Weingart (1998) has called the medialisation of science. The shift in the value orientation of the institution of science impacts on the trust granted to scientists, the public understanding of the nature and function of science and the reliability of scientific knowledge. With the media system – from research organisations through to news outlets – placing such a premium on newsworthiness and persuasion, science journalism risks ending up both less informative and less reliable.

There are some patterns indicative of this reinforcing of media logic that are hard to ignore. A language-oriented study by Taylor (2010) compared top British outlets in 1993 and 2005 and found more sensationalised and unspecified references to "the science" behind reported issues in the later period than the earlier one. Science journalists report less and less on the specifics of the science and how it is conducted, Taylor suggests, preferring instead to use shortcuts such as "the science behind", "the science of", "all the science", "the science bit", "the scientists warn" and so on. Information that used to be mentioned but is now much rarer includes research caveats and implications, researchers' methodological choices, the status of research as reviewed or replicated and the names of institutions behind particular studies (see Mellor, 2015, for similar observations). In place of these, journalists focus instead on recency ("a study published today") and sentiment ("scientists fear") (Taylor, 2010: p. 234).

Such generalisations are particularly significant as they can shore up editorial endorsement of political and economic policies or ideological positions without having to specify research institutions, funders or publishers; details that might reveal a more ambiguous situation. Sentence structures that repeatedly declare that "science proves" an issue or ones that use the "science argument" for a conclusion may be especially problematic. Such phrases tend to prime the readers to accept claims and construct consensus through an appeal to some "unspecified, autonomous, authoritative scientific entity" (Taylor, 2010: p. 236). The cumulative effects of this reliance on generalisation have the potential to influence public perceptions of the nature of scientific evidence and its role in debates about public policy.

The affective turn: Emotional engagement

As with undue newsworthiness and shortcuts, which represent almost anything as a breakthrough and simplify scientific argument, an overemphasis on emotion may also detract from the public's capacity for reasoned engagement with scientific developments. And yet, recruiting emotions for the sake of keeping public attention on science has never been more important. Here I return to the central issue of the tension between style and substance from the perspective of affectivity.

The media's affective turn has prompted scholarly interest in how emotions are catalysed in news reporting of topics that do not normally concern readers' lifeworlds (Ahmed, 2004; Lutz and Abu-Lughod, 1990). People are most likely to be emotionally invested in coverage of other people (often children), in news that relates to their culture and identity, or in health advice (Weitkamp, 2003). Thus, for example, one way to increase the audience's emotional investment in science is to relate news about routine lifestyle choices, diet, hygiene or exercise, to science by "medicalising" it (Bauer, 1998; Hallin et al., 2013). This is often done through specific jargon (e.g., "perspiration" rather than "sweating") and an artificial problem-solution compositional schema (Molek-Kozakowska, 2019).

When topics are relatively mundane, presenting a dramatic aspect – especially if related to injury or misfortune, however hypothetical and easily dismissible – is likely to increase engagement. For instance, an analysis of the coverage of cycling in the international press found that reporters focused on accidents much more than on health benefits (Bednarek and Caple, 2017). Similarly, Polish news coverage of the Covid-19 vaccine foregrounded incidental negative side-effects rather than public health advocacy and recommendations (Molek-Kozakowska and Struchkova, 2022). Such dramatic framings are not new – negative news, health scares and media panics have been selling newspapers for at least a century (McManus, 1994; Tomes, 2002).

However, not all news coverage evokes negative emotions. As the above list of news values indicates, journalists also value positivity. Positive appeals to curiosity, wonder and awe are commonly deployed by science communicators of all types (Davies, 2019; Silva Luna and Bering, 2020). Awe may be a particularly valuable mobilising emotion in science journalism, as it fosters curiosity, admiration, revelation and connection on top of promising entertainment. This type of affective appeal is common when the science being reported has little relation with readers' lifeworlds yet is still remarkable. For example, in the *New Scientist* article "How the secrets of ancient cuneiform texts are being revealed by AI", excitement and awe are constructed stylistically through metaphor (e.g., "unlocking new secrets") and contrastive frames ("today's" versus "ancient"):

> A photographer is meticulously capturing images of this writing, with lights positioned to highlight the indented etchings. This work is part of a revolution, one that is using today's computing power to bring this 5000-year-old record back to life and unlock new secrets of the world's first civilisation.
>
> (George, 2022)

Humour also plays a role in evoking positive emotional responses to science news. In the case of news that goes viral on social networks, where the users themselves are freer to select what to react to, stories presented with a dose of humour can offset the negativity of much of the news. For instance, a study of viewer engagement with scientific messages on Twitter found that messages containing linguistic humour (especially wordplay) were likely to attract more reactions, such as retweets, than were messages that related the same scientific fact but without humour (Yeo et al., 2020). Similar audience preferences are likely to apply to science journalism on other platforms. Indeed, popular science magazines make use of humour even when dealing with serious topics. Thus, in its report of a possible flaw in the data from a new space telescope, *New Scientist* uses irony to present the problem as a consequence of success: "The James Webb Space Telescope (JWST) is performing better than expected, and that might be a problem for some of the early results" (Crane, 2022). However, some styles of humour – such as sarcasm and satire – may be inconducive to evidence-based scientific literacy building, especially if used in decontextualised social media highlights. It is notable that this type of humour is absent from the *New Scientist* trending articles.

Anthropomorphism is an even more common affective strategy, as is evident from the *New Scientist* headlines, in which actions predicated on human motivation or communication are attributed to abstract entities: a chemical "allows" mice to live longer, cold exposure "hijacks" glucose stores or the Earth's infancy "tells us" things.

Emotional engagement can also be elicited by describing the human dimension of how science is done, as in the above description of the photographer working on the ancient cuneiform texts or when the article about the James Webb Space Telescope quotes an astronomer saying: "it was a bit of an 'astronomers at Christmas' scenario with everyone diving in to see what they could find" (Crane, 2022). Yet, while it can be inappropriate to offer just numbers and estimates of deaths, victims or patients in ways that objectify humans, a focus on individual experience can also diminish the newsworthiness of an event. *New Scientist*'s "40°C heatwave may have killed 1000 people in England and Wales" explains how the researcher produced the estimate but does not tell the personal story of anyone who died in the heatwave. Instead, the article juggles with ratios – "more than half of the deaths are expected to have been in people aged 85 and older" (Vaughan, 2022) – to underscore the significance of the heatwave through an appeal to Negativity and Superlativeness.

The affective turn – in its evocation of awe and its description of human endeavours – can easily result in the hyping of scientific developments. Science journalists, therefore, need not only to create interest but also to moderate public expectations. An example of how this is done can be found in the apparently factual headline "Blood substitute repairs damaged organs hours after heart stops", which goes on to explain that:

> The system could also potentially be used to help people who have had heart attacks or strokes – when blood supply to the heart or brain, respectively, is reduced – by perfusing either of these organs with the healing fluid. "If it is successful in resuscitating an organ which has suffered an otherwise fatal injury, then potentially this is very exciting," says Friend.
>
> (Wilson, 2022)

Here, emotions and attitudes ("successful", "exciting") are attributed to the researchers rather than being presented as the judgement of the journalist, and are moderated through the inclusion of the conditional "if" and the repeated "potentially". In walking this line between promise and caution, affective and ethical concerns usually get tangled together, as when this article hints at anxieties around clearing organs for transplantation following a fatal incident or when it later explicitly references ethical concerns about trying to "reverse death". It is important, then, for the science journalist to know how to calm emotions down, not only how to hype them up.

The affective turn thus returns us to the dilemma inherent in the systemic shift from "truth value" to "news value" (Weingart, 2022). In a media system where science journalists have to compete for attention, managing popularity means having to exploit emotional investments (Broks, 2006; Duan et al., 2021; Molek-Kozakowska, 2017). Media scholars who try to explain virality note that it is often driven by mere physiology. In a marketing-oriented study, Berger and Milkman (2012) find confirmation for a claim that content that evokes high-arousal positive emotions (e.g., awe) or negative emotions (e.g., anger or anxiety) is more viral, whereas content that evokes low-arousal, or deactivating, emotions (e.g., sadness) is less viral. However, the relationship between emotion and social transmission is complex and cannot be reduced to physiology alone. As studies on the coverage of therapeutic cloning, CRISPR, geoengineering and AI have shown, an emotional "dialectic between hope and fear" (Jensen, 2012: p. 40) is yet another strategy to keep loyal readers coming back for more science news.

The narrative turn: Science storytelling

Scientists and science communicators are increasingly being urged to recast science in the form of stories or narratives. Narratives are the textual representations of events through plotlines and interpretative frames that attribute specific meanings, causal relations and values to actors, objects, actions and processes. These are deemed to be more engaging, easier to process and more persuasive than other compositional choices (Dahlstrom, 2014). Of course, stories have long been the stock in trade of the journalist and science journalists, too, are in the business of telling stories.

Journalistic storytelling is most frequently used to personalise experiences or activities that in turn can make science seem relevant. For instance, the *New Scientist* article "We are finally starting to understand brain fog and how to treat it" likely trended not because of its high-profile discoveries, but because of the everyday relevance of its message. The text does not announce a breakthrough, but rather puts the known, yet scattered, facts into perspective. The use of a framing narrative provides a coherent and ordered structure in which to place the confusing array of long-term health impacts of Covid-19. The article starts with a short anecdote:

> Courtney Shukis was looking forward to lunch: she had just recovered from covid-19 and was glad to be meeting her friends again. Before leaving her home in Plano, Texas, she checked the calendar, making a mental note of the restaurant and when to meet. "But instead of going there, I got in my car and drove to a completely different place," she recalls. "I sat at the table for half an hour, looking at my phone, wondering where everyone was. My brain fog was really bad."
>
> (Sukel, 2022)

The reference to a named individual, the vivid details of the setting and the quick-paced sequence of actions (with recognisable causal relations between them) are followed by a climactic situation to be resolved at the end of the article. Direct quotes and attributed thoughts add a first-person point of view to the narration. This allows the story to be imbued with emotion ("look forward to", "wonder") and evaluation ("really bad"), thereby building the motivation to pursue the text's multiple points and to reflect on its implications.

The reasons for incorporating stories into science journalism are many. The *New Scientist* article "Is our solar system a cosmic oddity? Evidence from exoplanets says yes" plays a humorous game with the loyal reader, who will be able to appreciate the narrative conceit:

> Once upon a time, there was a solar system. In it lived four small rocky planets called Mercury, Venus, Earth and Mars. Four big gassy planets lived there too: Jupiter, Saturn, Uranus and Neptune. The four small planets lived close to the sun because it was very hot there, and everything else had been blasted away, leaving only rocks to make planets. But further out it was colder and there was lots of ice around, so the planets there grew into great big gassy giants.
>
> (Clark, 2021)

Here, a recurrent ritualistic element such as anthropomorphism enables the science journalist to endow planetary evolution with a narrative structure, aligning astronomy with readers' intellectual and emotional needs even while gently mocking the recourse to this stylistic device.

Reclaiming science from jargon-laden academia and accommodating it to the conventions of popular culture is a challenging stylistic enterprise. Storytelling is often embraced as a strategy that can open up and democratise science by making it approachable. Yet the term "media narrative" is also used to describe the more insidious practice of a media outlet creating and propagating a specific storyline through which incoming news items are filtered and along which they are subsequently interpreted. Hence, narrative is treated as a rhetorical device as well as a compositional form, and as such, it can function as a tool to exercise power, no matter if the story told is scientific, historical or fictional (Steinhauer, 2022). Journalists construct narratives out of events, giving them labels – for example, 'a discovery', 'an official inquiry', 'a promotion' – and in so doing offering them for specific interpretations. These interpretations are usually compatible with their outlet's editorial line and interests.

So compelling is narrative that it risks locking in interpretative structures that are inconsistent with the scientific evidence or that misinterpret scientific uncertainty. The unbounded messiness of real life is marshalled into a single storyline with an identifiable protagonist experiencing singular events. For example, Wong and Claypool (2021) discuss the widespread desire to close down the proliferation of narratives during the Covid-19 pandemic despite being "confronted with a thoroughly postmodern disease that refuses to be contained to a single plot line". For Wong and Claypool what was remarkable about the pandemic was "our insistence to find a single explanatory narrative to make sense of it." Competing and conflicting mask-wearing narratives, each claiming to be *the* correct narrative, were, they argue, a consequence of social and political, not just biological, conditions. As these pandemic narratives suggest, although science journalists deploy narratives as a captivating sense-making structure, they can also function as an articulation of power.

Conclusion

Stylistic differences between the journalistic voice and the scientific voice are the consequence of the political, economic and cultural conditions in which news reporting is situated. Taken together, the three stylistic turns discussed in this chapter prioritise attention-gathering over explanation. They enable journalists to reach a public audience but at the same time they risk obscuring, or even displacing, the information on which science news is based.

Journalists' stylistic choices project newsworthiness onto scientific events that are not innately newsworthy, drawing attention to developments that would otherwise appear dull or irrelevant. Stylistic choices are also often steered towards inducing and managing emotion. Affectivity can be evoked by highlighting personal relevance, making the routine seem extraordinary, dramatising mundane events and through the use of humour. Compositional aspects of style also play a part. Representing a science news item in the form of a story with sequential patterns and easily recognisable characters can help attract readers and align their cognitive and emotional reactions. These three turns – attentional, affective and narrative – are usually mutually reinforcing. When taken to the extreme, their combined focus on engagement through stylisation and contrivance, rather than on enhancing science literacy, can compromise the ability of science journalism to share scientific information.

Finding a socially responsible journalistic voice demands a stylistic repertoire that can navigate these tensions between truth value and news value. This challenge is exacerbated by the research culture in which science is produced. Science communication scholar Martin Bauer argues that the commercialisation of research is propelling science communication further towards science PR, marketing and virality. He notes the resulting proliferation of hype, even "bullshit", in science journalism and calls for increasingly critical science-literate readers: "A sceptical public that is not awestruck by the new displays of science is a necessity for knowledge societies" (2008: p. 23).

Bauer seems to be opting for the audience's "inoculation" to the shifts from truth value to news value, yet readers are not the only ones who need to be vigilant of the excesses of PR-driven science dissemination. Professional media organisations, social movements and public regulators all also have a role to play. Reliable science information that impinges on well-being, health, prosperity or life should not be dispensed like any other commodity – either given away in free colourful samples or available only for those who can afford it. Likewise, science journalists should continue on their mission to engage the public with informative science. To do so, they need to attend carefully to their style.

Note

1 To select examples of typical journalistic practice, the sampling and downsizing for this case study was carried out according to the so-called 'constructed week' protocol, whereby the five top trending articles were harvested from the outlets' websites on a Sunday in the first week, on a Saturday in the second week, etc., from 11 June to 15 August 2022. Most trending articles are recently published. However, a few in the sample dated back as much as five years as they had been hyperlinked to newer articles and drew enough readers to meet the criteria of the magazine's trending algorithm.

References

Ahmed, S., 2004. *The Cultural Politics of Emotion*. London: Routledge.

Amsen, E., 2022. Don't Look Up points out science communication challenges. *Forbes*, https://www.forbes.com/sites/evaamsen/2022/01/06/dont-look-up-points-out-science-communication-challenges/?sh=aab059549ae0.

Angelone, S., 2022. *Don't Look Up*: Science communication revisited. *Science Communication*, 44(3), pp. 375–382.

Badenschier, F. and Wormer, H., 2012. Issue selection in science journalism: Towards a special theory of news values for science news? *Sociology of the Sciences Yearbook: The Sciences' Media Connection – Public Communication and Its Repercussions*, 28, pp. 59–85.

Bauer, M.W., 1998. The medicalization of science news: From the rocket-scalpel to the gene-meteorite complex. *Social Science Information*, 37(4), pp. 731–751.

Bauer, M.W., 2008. Paradigm change for science communication: Commercial science needs a critical public. In: Cheng, D. et al (ed.), *Communicating Science in Social Contexts*. New York: Springer, pp. 7–25.

Bednarek, M., 2016. Inventory of linguistic devices constructing newsworthiness. *Discursive News Values Analysis*, https://www.newsvaluesanalysis.com/what-is-dnva/.

Bednarek, M. and Caple, H., 2014. Why do news values matter? Towards a new methodological framework for analyzing news discourse in critical discourse analysis and beyond. *Discourse and Society*, 25(2), pp. 135–158.

Bednarek, M. and Caple, H., 2017. *The Discourse of News Values*. Oxford: Oxford University Press.

Bell, A., 1991. *The Language of News Media*. Oxford: Blackwell.

Berger, J. and Milkman, K.L., 2012. What makes online content viral? *Journal of Marketing Research*, 49(2), https://doi.org/10.1509/jmr.10.0353.

Blom, J.N. and Hansen, K.R., 2015. Click bait: Forward-reference as lure in online news headlines. *Journal of Pragmatics*, 76, pp. 87–100.

Broks, P., 2006. *Understanding Popular Science*. Maidenhead: Open University Press.

Clark, S., 2021. Is our solar system a cosmic oddity? Evidence from exoplanets says yes. *New Scientist*, 1 Dec., https://www.newscientist.com/article/mg25233630-600-is-our-solar-system-a-cosmic-oddity-evidence-from-exoplanets-says-yes/.

Dahlstrom, M.F., 2014. Using narratives and storytelling to communicate science with nonexpert audiences. *Proceedings of the National Academy of Sciences*, 111(4), pp. 13614–13620.

Davies, S.R., 2019. Science communication as emotion work: Negotiating curiosity and wonder at a science festival. *Science as Culture*, 28(4), pp. 538–561.

Duan, R., Hepworth, K.J., Ormerod, K.J. and Canon, C., 2021. Promoting concern for climate change: A study of wildfire photographs using q methodology. *Science Communication*, 43(5), pp. 624–650.

Entradas, M. and Bauer, M.W., (eds.). 2022. *Public Communication of Research Universities: 'Arms Race' for Visibility or Science Substance?* New York: Routledge.

Fahnestock, J., 1993. Accommodating science: The rhetorical life of scientific facts. In: McRae, M.W. (ed.), *The Literature of Science*. Athens: University of Georgia Press, pp. 17–36.

Fahy, D., 2022. Caricatures and omissions: Representations of the news media in 'Don't Look Up'. *Journal of Science Communication*, 21(5), p. C07. https://doi.org/10.22323/2.21050307.

Galtung, J. and Ruge, M.H., 1965. The structure of foreign news: The presentation of the Congo, Cuba and Cyprus crises in four Norwegian newspapers. *Journal of Peace Research*, 2(1), pp. 64–90.

George, A., 2022. How the secrets of ancient cuneiform texts are being revealed by AI. *New Scientist*, 3 Aug., https://www.newscientist.com/article/mg25533981-400-how-the-secrets-of-ancient-cuneiform-texts-are-being-revealed-by-ai/.

Guenther, L., and Ruhrmann, G., 2013. Science journalists' selection criteria and depiction of nanotechnology in German media. *Journal of Science Communication*, 12(3), pp. 1–17.

Hallin, D.C., Brandt, M. and Briggs, C.L., 2013. Biomedicalization and the public sphere: Newspaper coverage of health and medicine, 1960s-2000s. *Social Science & Medicine*, 96, pp. 121–128.

Jensen, E., 2012. Scientific sensationalism in American and British press coverage of therapeutic cloning. *Journalism & Mass Communication Quarterly*, 89(1), pp. 40–54.

Jung, A.K., Stieglitz, S., Kissmer, T., Mirbabaie, M. and Kroll, T., 2022. Click me…! The influence of clickbait on user engagement in social media and the role of digital nudging. *PLOS One*, 17(6), p. e0266743. https://doi.org/10.1371/journal.pone.0266743.

Lutz, C.A. and Abu-Lughod, L., (eds.). 1990. *Language and the Politics of Emotion.* Cambridge: Cambridge University Press.

McManus, J., 1994. *Market-Driven Journalism.* Thousand Oaks: Sage.

Mellor, F., 2015. Non-news values in science journalism. In: Rappert, B. and Balmer, B. (eds.), *Absence in Science, Security and Policy: From Research Agendas to Global Strategy.* Basingstoke: Palgrave Macmillan, pp. 93–113.

Molek-Kozakowska, K., 2013. Towards a pragma-linguistic framework for the study of sensationalism in news headlines. *Discourse & Communication*, 7(2), pp. 173–197.

Molek-Kozakowska, K., 2016. Stylistic analysis of headlines in science journalism: A case study of *New Scientist. Public Understanding of Science*, 26(8), pp. 894–907.

Molek-Kozakowska, K., 2017. Communicating environmental science beyond academia: Stylistic patterns of newsworthiness in popular science journalism. *Discourse & Communication*, 11(1), pp. 69–88.

Molek-Kozakowska, K., 2019. Making biosciences visible for popular consumption: Approaching image-text relations in Newscientist.com through a critical multimodal analysis. *Qualitative Inquiry*, 25(4), pp. 379–392.

Molek-Kozakowska, K. and Struchkova, S., 2022. Communicating risks of an anti-COVID-19 vaccine in Poland: A comparative case study of content, style and advocacy of three media outlets. In: Hohaus, P. (ed.), *Science Communication in Times of Crisis.* Amsterdam: Benjamins, pp. 143–168.

New Scientist, 2023. About New Scientist magazine. *New Scientist*, https://www.newscientist.com/about/.

Olson, R., 2015. *Houston, We Have a Narrative: Why Science Needs Story.* Chicago: University of Chicago Press.

Rutkin, A., 2016. How Minecraft is helping children with autism make new friends. *New Scientist*, 27 Apr., https://www.newscientist.com/article/mg23030713-100-how-is-helping-children-with-autism-make-new-friends/.

Silva Luna, D., and Bering, J.M., 2020. The construction of awe in science communication. *Public Understanding of Science*, 30(1), pp. 2–15.

Sparkes, M., 2022. DeepMind's protein-folding AI cracks biology's biggest problem. *New Scientist*, 28 Jul., https://www.newscientist.com/article/2330866-deepminds-protein-folding-ai-cracks-biologys-biggest-problem/.

Steenstrup-Duch, A., 2022. Don't Look Up: Can effective science communication save the world? *#SINTEFblog*, https://blog.sintef.com/sintefenergy/dont-look-up-can-effective-science-communication-save-the-world/.

Steinhauer, J., 2022. *History, Disrupted: How Social Media and the World Wide Web Have Changed the Past.* Cham: Palgrave Macmillan.

Sukel, K., 2022. We are finally starting to understand brain fog and how to treat it. *New Scientist*, 8 Jun., https://www.newscientist.com/article/mg25433902-300-we-are-finally-starting-to-understand-brain-fog-and-how-to-treat-it.

Taylor, C., 2010. Science in the news: A diachronic perspective. *Corpora*, 5(2), pp. 221–250.

Tomes, N., 2002. Epidemic entertainments. Disease and popular culture in early-twentieth-century America. *American Literary History*, 14(4), pp. 625–652.

UniProt Consortium, 2022. *UniProt: the Universal Protein Knowledgebase*, https://www.uniprot.org/.

Vaughan, A., 2022. 40°C heatwave may have killed 1000 people in England and Wales, *New Scientist*, 28 Jul., https://institutions-newscientist-com.iclibezp1.cc.ic.ac.uk/article/2331349-40c-heatwave-may-have-killed-1000-people-in-england-and-wales/.

Weingart, P., 1998. Science and the media. *Research Policy*, 27(8), pp. 869–879.

Weingart, P., 2022. Preface. In: Entradas, M. and Bauer, M.W. (eds.), *Public Communication of Research Universities: 'Arms Race' for Visibility or Science Substance?* New York: Routledge, pp. xxi–xxii.

Weitkamp, E., 2003. British newspapers privilege health and medicine topics over other science news. *Public Relations Review*, 29(3), pp. 321–333.

Wilson, C., 2022. Blood substitute repairs damaged organs hours after heart stops. *New Scientist*, 3 Aug., https://www.newscientist.com/article/2331983-blood-substitute-repairs-damaged-organs-hours-after-heart-stops/.

Wong, J., and Claypool, E., 2021. Narratives, masks and COVID-19: A qualitative reflection. *Qualitative Social Work*, 20(1-2), pp. 206–213.

Yeo, S.K., Yi-Fan Su, L., Cacciatore, M., McKasy, M. and Qian, S., 2020. Predicting intentions to engage with scientific messages on twitter: The roles of mirth and need for humor. *Science Communication*, 42(4), pp. 481–507.

10

ETHICAL COMPROMISE IN NARRATIVE SCIENCE JOURNALISM

The Immortal Life of Henrietta Lacks

Lauren Kilian

When Rebecca Skloot's first book appeared in 2009, it invited readers to delve into what might have seemed like an esoteric topic: a development in biomedical research methods and the woman born in rural Virginia who made that development possible. In the time since, *The Immortal Life of Henrietta Lacks* has brought widespread attention both to Henrietta Lacks's short life and to the "immortal life" her cells have lived by becoming one of the most robust lines of immortal human cells ever known and the first to be "grown in a laboratory" (Skloot, 2010: p. 4). The book was to have an impact as far-reaching as that of the cells it is about, becoming a rare example of book-length science journalism that has topped the bestseller lists and been adapted as a film. It has won multiple awards, educated the public about the use of human cells in laboratories, generated discussion about racial disparities in access to health insurance, exposed the lack of control that most Americans have over tissue samples removed from their bodies and become part of a movement that calls for changes in how consent is obtained in medicine and research.

Immortal Life's success stems from its use of narrative: it employs narrative productively and responsibly while still being provocative and impactful. The book's reception and impact are a result of narrative strategies that elicit investment from the reader. But that investment comes at a price. Ironically, for a book whose aim is to expose a breach of ethics, Skloot's writing entailed its own ethical compromises. This chapter examines the narrative strategies of *Immortal Life* for what they reveal about the ethical choices entailed in narrative journalism.

Skloot's approach was to weave together several distinct storylines. In one, Skloot traces the origin of the HeLa cell line to the tissue sample that Lacks's

DOI: 10.4324/9781003326724-14

doctor removed while he was treating Lacks for cervical cancer. The fact that he did so without Lacks's knowledge, let alone her consent, becomes one of Skloot's central concerns. A second strand of the narrative follows the journey of that tissue sample. As Skloot puts it, once in the laboratory, the cells "reproduced an entire generation every twenty-four hours, and they never stopped" (Skloot, 2010: p. 4). The cell line has since gone on to be instrumental in such ground-breaking developments as the polio vaccine, genetics, cancer treatment, in vitro fertilisation and countless other areas of research. The book's success, however, can be traced to a third storyline: Skloot's own journey of discovery and her interactions with Henrietta Lacks's family. This serves to collapse the distance between the journalist, her subjects and her readers, ensuring that a potentially obscure topic – scientific research and medical ethics – is rendered accessible and relatable. Yet it also creates ethical dilemmas for the author.

Like fiction, narrative non-fiction establishes a relationship between author and reader, which in turn mediates the relationship between reader and subject. According to narrative theorist Wayne Booth, all narratives offer "a cry of invitation: 'Join me, join me, because if you do, you will receive something that no other story can give you in quite the same way, something that will ...'" – a statement Booth leaves incomplete to represent the countless ways narrative paths can "diverge" from that moment of invitation (Booth, 1988: p. 174). By telling immersive stories, narrative journalism leads the reader down some paths and, in so doing, directs their attention towards certain events and characters and away from others. These diverging paths bring with them choices and compromises. As Booth emphasises, these are ethical in nature. Booth outlines the different possible author-reader dynamics that can result, including "reciprocity", in which readers and authors are positioned as "equals", and "hierarchy", in which readers are either "inferiors or superiors" (Booth, 1988: p. 184).

Skloot's decision to present a first-person account of her own journey invites the reader to join her in a reciprocal relationship. Yet, in writing herself into the story, Skloot was to forever link her experiences as a journalist with the story of the cells and the family she was reporting on, ultimately raising questions about whose story it is to tell. In what follows, I consider how Skloot pivots her complex and multi-stranded narrative around her own presence as a character in the story. *Immortal Life* amply demonstrates that the narrative immersion of book-length journalism can succeed in engaging readers with difficult material but, I argue, this approach is founded on an ethical misdirection. In attempting to recover the lives and experiences of a Black woman and her family, Skloot refracts their story through the outsider subjectivity of a white journalist. The ironic tension at the heart of Skloot's book is that in order to uncover the ethical betrayal enacted by Lacks's doctors and the scientists who cultured her cells, Skloot, like all literary journalists, must choose between conflicting ethical impulses and must submit these ethical concerns to the demands of narrative.

Narrative journalism and the first-person narrator

We can trace the modern form of narrative journalism – also known as literary journalism – to the emergence of the New Journalism movement of the 1960s and 1970s. The work of American journalists such as Joan Didion, Norman Mailer, Hunter S. Thompson and Tom Wolfe demonstrated that non-fiction books could immerse readers in social worlds far removed from their own. Tom Wolfe's *New Journalism* anthology, co-edited with Edward Warren Johnson and published in 1973, helped identify the form as a distinct genre within journalism, one that made liberal use of storytelling techniques akin to those used in fiction, sometimes to the extent of blurring the traditional boundaries between fiction and non-fiction. Indeed, as Kobe van Krieken and José Sanders highlight in a review of the scholarship on narrative journalism, the genre represents a deliberate disruption of these boundaries, creating tensions "between ethics and aesthetics, fact and fiction, and objectivity and subjectivity" (van Krieken and Sanders, 2019: p. 1394).

Some scholars go further and claim that book-length narratives reveal objectivity to be an unattainable ideal. Lisa Phillips (2019), for instance, argues that literary journalism is characterised by an inevitable subjectivity, reflected most explicitly in the common use of the first person. As journalist Janet Malcolm famously put it: "the 'I' character in journalism is almost pure invention" (1990: p. 22). Furthermore, argues communications scholar David Eason, the literary journalist must adopt an inherently paradoxical tone, as this "I" must be "simultaneously near and far" (1984: pp. 55-56). While the common notion of the journalist as a passive witness of events implies detachment, the literary journalist must minimise this distance from their subjects to elicit the reader's compassion for the subjects of the story. To achieve this, the journalist will likely cultivate a personal relationship with their sources. In the introduction to his *New Journalism* anthology, Tom Wolfe warned that this could lead especially sensitive journalists to "become stricken with a sense of guilt, responsibility, obligation" (Wolfe and Johnson, 1973: p. 51); the implication being that, in Wolfe's opinion, good journalists should feel no such qualms.

As this suggests, narrative journalism raises significant ethical issues. Indeed, in *The Ethics of the Story*, David Craig suggests that the ethical choices that all journalists must make are "particularly evident in narrative writing" (2006: p. 2). For the literary journalist, the imbalance in power between journalist and subject, already present in much journalism, is made more acute by their insertion into their subjects' lives over a long period. As Phillips explains:

"Journalists decide when to show up and when not to, exercising control over their own presence in ways their subjects cannot. This element of control remains no matter how central journalists are to their narratives and how much they disclose about themselves" (2019: p. 393).

Similarly, Matthew Ricketson (2021) proposes that narrative journalism, especially at book length, presents unique ethical concerns. How do narrative journalists reconstruct scenes created from multiple sources and information gathered over months or even years? Can such scenes, with dialogue and detail, be an accurate record of the events they describe? Further, how can the journalist develop the intimacy that is imperative to narrative journalism while also maintaining a professional distance from their sources? And how can the journalist craft compelling and personalised portraits of their principal sources at the same time as minimising harm to them? Transparency is often cited as the answer to these ethical quandaries. As will be discussed, the genre's characterising feature of subjectivity is thus configured as its ethical guarantor, even, at the same time as it promotes a limited perspective.

The pairing of narrative journalism and scientific subject matter arguably makes these ethical concerns even more challenging. Added to the concerns surrounding appropriation, attribution, exploitation and sensitivity that arise with any narrative journalism, science journalism is faced with making accessible to a general audience subjects that Craig describes as having "intellectual complexity" (2006: p. 2). While similar challenges arise for reporters in other fields, Craig argues that fields such as science, medicine and technology, with dense content and terminology that may be unfamiliar to the lay public, present unique obstacles (p. 74).

The view of science as revealing unambiguous objective truths about the world potentially further aggravates these ethical tensions. If science is usually perceived to speak with one authoritative voice, the voice of the journalist may come across as unreliable. Journalists' books about science enter a popular book market dominated by scientist authors. The ethical journalist, writing themselves into the story in order to be transparent about their subjectivity, risks appearing untrustworthy when juxtaposed with subjects – and other authors – who claim a higher level of objectivity. At the other extreme, scientific subject matter might reinforce what Malcolm called the "over reliable narrator" (1990: p. 22), who stands apart at a dispassionate distance. Readers may be unwilling to recognise the journalistic "I" as a literary construct if the journalist's role is to report on a story world so closely associated with objectivity.

In *Immortal Life*, Skloot's authority to narrate the science was not the only challenge she needed to navigate. Most important for this discussion is the fact that Skloot is a white author chronicling the racialised experiences of a community to which she does not belong, something she is explicit about in the prologue to the book (p. 7). Skloot's remove from her subjects introduces the ethical dilemmas that this chapter explores. The book is a work of advocacy, with the ambitious aim to explore the intersection of "science, ethics, race, and class" (p. 317). To spotlight the underreported story of the woman and family tied to the history of the HeLa cell line, to chronicle the impact of HeLa, to advocate for the importance of informed consent in scientific and medical research and to achieve these worthy pursuits with the undeniably successful outcomes that she does, Skloot had to find a way of collapsing the distance between herself and her subjects, and to invite readers to join her there.

From singular minds to a multitude of singular stories

Skloot's approach appears, at first glance, to be a conventional one: she crafts a narrative in which real-life figures are cast as characters in a dramatic sequence of events. Media scholars Matthew Nisbet and Declan Fahy attribute the success of such a model to the crafting of a "dramatic plot" that imposes a "specific temporal order" on the real events and guides the audience towards "moral lessons" by the narrative's conclusion (2013: p. 3). The book's overall arc traces Skloot's early hunger for knowledge of the woman behind the HeLa cell line and delves into the histories of the figures she encounters along the way, transforming the raw material that she has amassed during a decade of reporting into a work that is routinely described as resembling a novel (e.g., Lasso, 2011; Roston, 2010; Seaman, 2009).

However, what sets *Immortal Life* apart from previous popular science narratives and long-form science journalism is who she casts as the primary characters in the narrative. Skloot spends comparatively less time with scientists than with non-scientists. The book is replete with figures who are very much outside the world of science. In fact, only one scientist figures prominently in *Immortal Life*: George Gey, the researcher in whose lab the HeLa cell line was created. Although Skloot presents Gey as an enthusiastic developer of the cell line, she bestows most of the glory on the properties of the cells themselves. She also notes that the success of the cell line is something Gey merely happens upon – and shares with others in the spirit of collaboration – rather than something that he ingeniously devises himself.

Rather than focus on the developments HeLa initiates or the scientists responsible for those developments, Skloot instead focuses on the people on whom they have had a lasting impact. For comparison among other successful popular science books that take a narrative approach, consider Dava Sobel's *Longitude*, Sylvia Nasar's *A Beautiful Mind* and Andrew Hodges's *Alan Turing: The Enigma*, which portray John Harrison, John Nash and Alan Turing, respectively, as men with singular minds set apart from society by their genius. While many of the people Skloot portrays may also be at the margins of society, it is not because they are celebrated as geniuses, but because they are Black Americans from rural areas who lack socio-economic power.

The combined result of Skloot's subject matter and the attributes of the people she profiles is that rather than providing insight into the inner workings of a scientific mind, Skloot tells a less sensational story that resists the 'lone genius' trope. In doing so, she finds a large audience who has a good chance of relating to the figures in the book, as the average reader will likely identify with people who don't know much about what happens in a lab. The injustices recounted will therefore resonate more profoundly.

The connection that Skloot forges begins with Henrietta Lacks. In Skloot's depiction, Henrietta looms as large as the HeLa cells, with as impressive a legacy. Skloot's portrait of Lacks's life moves from one vivid scene to the next, from Henrietta working in the tobacco fields as a child (p. 20), to sneaking away from her house as a young woman to go dancing (p. 42), to telling her cousin at the top of a Ferris wheel that she has cancer (p. 46). Despite persistent optimism, her life never ceased to be difficult: her husband was unfaithful and she was profoundly impacted by her eldest daughter's epilepsy and institutionalisation. When Henrietta dies, Skloot recounts, a cousin calls out: "let that poor woman rest" (p. 91). The vivid imagery and extensive interviews create a full and compelling portrait.

Skloot is also interested in the lives of those Henrietta leaves behind – her husband, siblings, cousins and children. Early in her investigations, Skloot is limited to talking to cousins or others who knew Henrietta in her tiny hometown or after moving to Baltimore. Yet with each new set of characters she introduces, Skloot moves closer to Henrietta – and to Deborah, Henrietta's grown daughter. As nearly every reviewer of the book notes, despite limited education, Deborah has a perceptive grasp on the family's complex injustices – ranging from being kept in the dark about Henrietta's cells for decades to being victims of a con man who eventually sued the family. Over the course of her life, Deborah endured abuse from the woman who cared for her after her mother died, sexual advances from a cousin as a teenager and strife with her husband and sons later in life. By the time Skloot met her, Deborah was both desperate for information about her mother and epileptic sister and drained by many ordeals. Skloot portrays these experiences as contributing to skittishness and distrust while also leaving Deborah warmly enthusiastic toward anyone who can help her in her search. Eventually, Deborah emerges as a partner in Skloot's quest. During their years of investigation and conversation, Deborah is at times trusting and eager, and at other times paranoid and guarded. In some moments, Deborah is a Virgil, leading Skloot through the underworld of her family's life. In other moments, she operates as a Watson to Skloot's Sherlock Holmes, with Skloot explaining what they uncover about Henrietta's cells.

Skloot takes up the story of the Lacks family as Henrietta's own life is coming to an end. At this point, the 'lives' of the cells also become a subject in Skloot's narrative. The reader watches as the tissue sample extracted from

Henrietta's cervix cheats fate and, unlike every previous sample in Gey's lab, survives indefinitely in culture. Skloot recounts how Gey shares vials of the HeLa cells with fellow researchers: first locally, then nationally and, ultimately, around the world. Eventually, the cells even make their way into space. Readers are invited to view the HeLa cells anthropomorphically: the cells cast as characters in their own adventure story, travelling the world and beyond to save humanity.

While granting inanimate objects agency is not uncommon in science journalism, what is distinctive about Skloot's approach is that in *Immortal Life*, this frequently takes the place of explanation. Science journalists are often expected to perform the role of 'explainer'; and putting her knowledge of biology on display would have been an option for Skloot, who has an undergraduate degree in biological sciences. Yet, there are many moments where Skloot has an opportunity to explain something but does not do so. For instance, when she asks Henrietta's son Lawrence if he knows what a cell is and the answer is "Not really" (p. 162), Skloot draws a diagram of a cell on the spot. Although she narrates her sketching of the diagram, she avoids sharing her explanation of it with the reader. In Booth's terms, rather than construct a hierarchical relationship with the reader by adopting the role of expert explainer, Skloot strives instead to establish a relationship of reciprocity; a relationship she might jeopardise were she to show off her expertise by sharing more of her knowledge about the cells. The story that Skloot tells – and the story that was to sell so well – is that she is on the reader's side.

The journalist as a character

Skloot achieves what she does with *Immortal Life* – popular and critical success and a role for the book in educational and policy contexts – by providing multiple entry points for the book's readers to invest in the story and to align themselves with the figures it portrays. Declining the role of expert and shifting the traditional focus of long-form science journalism from the actions of scientists to the people those actions impact are two of the means through which she accomplishes this. Another is by writing of herself as a character in the narrative. On her website, Skloot explains that she had resisted this for a long time, but eventually realised that it would be "dishonest" not to do so: it was, she says, a form of "disclosure" (Skloot, n.d.). Lindsay Morton (2014) identifies this decision as part of an "extensive network of disclosure transparency measures" through which Skloot attempts to signal her own ethical integrity as a journalist. Yet it raises new ethical problems in turn. Establishing herself as a character converts – but does not remove – the ethical question posed by her narrative: an epistemic issue, "Whose knowledge is this?", becomes a narratival one, "Whose story is this?" (Morton, 2014: p. 250).

Skloot places herself within the story from the very first page of the book: "There's a photo on my wall of a woman I've never met, its left corner torn and patched together with tape" (p. 1). Inviting readers into her home and sharing her lack of knowledge, Skloot immediately establishes herself as open and trustworthy. At the same time, the passage establishes intrigue: who is the woman?, why does Skloot have her picture?, why is it torn? By the end of the prologue, we have heard an account of the first time Skloot learned the name Henrietta Lacks in a biology class, we have followed Skloot home from school to witness her search for an encyclopaedia only to find nothing on Lacks and we have learned that in the course of the ten years it took to write the book, Skloot eventually met Deborah Lacks, the daughter of the woman in the picture. Before the story proper has opened, Skloot has tethered herself to it, crafting the 'reciprocity' that Booth finds when reader and author are equals.

This device of inserting herself into the story resurfaces continually, so that as Skloot weaves between biographies of members of the Lacks family and the biography of the HeLa cell line, the reader also witnesses a behind-the-scenes narrative of the journalist's search for the raw information from which she constructed the narrative. Consider Skloot's description of her arrival in Turner Station, the town to which Henrietta moved as a young woman. The scene includes not just her interview with town librarian Courtney Speed, who had been trying to start a museum about Henrietta, but the story of Skloot's journey to the interview. She describes circling the town "so many times" that a group of men "started waving at me with each pass" (p. 70). In a significant change of perspective, she then reverses the narrative gaze: "The people of Turner Station just watched me, smiling and shaking their heads like, *What's that young white girl doing driving around in circles?*" (p. 70, Skloot's italics). As with her pulling back from the role of explainer, this upending of describer and described might appear to surrender a crucial attribute of the first-person narrator as the filtering presence within the story. Yet Skloot retains control, passing off her own speculation as the thoughts of the townsfolk.

In another scene, Skloot draws attention to her identity as a writer, but again in a way that simultaneously diminishes and reinforces her role. She discovers she is staying in the same motel where Michael Rogers, a reporter who covered the story in 1976 for *Rolling Stone*, had stayed. She sees the same sign for 'BROMOSELZER' out the window that Rogers described seeing and uses the same phone book that Rogers describes using. She is not, then, the first reporter to track down the Lacks family. Yet in acknowledging this, Skloot establishes a pedigree: she is descended from predecessors who have followed this same path but from whom she ultimately distinguishes herself.

A key way in which Skloot establishes that distinction is in chronicling her attempts to gain the trust of the Lacks family. By the book's end, we witness an intimate exchange between Deborah – who by this point is ground down by the process of uncovering details of her mother's and sister's lives – and her cousin Gary Lacks. Skloot's presence at the scene demonstrates how ingrained in the family's life she has become, as she describes Gary lifting Deborah's burden of stress by pronouncing the cells excised from Deborah and placed on the shoulders of Skloot. The scene, like many others, does not advance the story of the cells or of Henrietta – nominally the purpose of the book – but functions instead as evidence that Skloot has earned the family's trust and, by extension, is worthy of the reader's trust. By demonstrating the family's apparent approval, Skloot once again collapses the role of describer and described, while maintaining her purchase on her position as author: if the family accepts her, she can legitimately act as both a fellow character in, and a storyteller of, their story.

Skloot has good reason to want to demonstrate her trustworthiness. As her relationship with the family grows more intimate, Skloot must confront the family's understandable weariness of outsiders who have written about the HeLa phenomenon before and treated the family as 'interesting colour', or have conducted further testing of blood samples from Lacks's children without informed consent. Despite the access she gained and the friendships she forged, Skloot did not fully gain the trust of all members of the Lacks family. Henrietta's oldest child, Lawrence Lacks, has publicly expressed his belief that Skloot's work – and the film adaptation of it – engages in the same kind of exploitation that he feels the family experienced at the hands of the medical research community (Giorgis, 2017).

Skloot attempts to neutralise this ambiguity about her position early in the book. In her encounter with Roman Patillo, the doctor who initially puts her in touch with the Lackses, Patillo first demands that she demonstrate she is informed enough to merit his help. He explains that he is wary of the family being exploited and asks Skloot about her knowledge of how the Black community has experienced abuses of medical ethics. Skloot launches into a description of the Tuskegee Syphilis Study and the "so-called Mississippi Appendectomies, unnecessary hysterectomies performed on poor black women to stop them from reproducing, and to give young doctors a chance to practice the procedure", as well as "the lack of funding for research into sickle-cell anaemia, a disease that affected blacks almost exclusively" (p. 50). The fact that Skloot includes the full record of her description of these events and that she describes the experience as like "giving a report in history class", makes the scene performative. She proves her authority and competency not only to Patillo, but to the reader.

By her own account, Skloot was reluctant to write herself into the story and aware of the ethical implications of such a move. In an interview a few months after the book's publication, Skloot explained: "I spent so much time fighting against being in the book, thinking, 'It's not my story, it's their story. It's not about me.' And I was right, it's not. It's just that I became a character in their story" (Pitzer, 2010). She therefore adopted strict criteria for which aspects of her own story to include:

> You don't learn anything about my backstory unless it's relevant to Deborah or to the family's story in some way. [...] You don't really learn that much about me as a character outside of their story. That was what I constantly had in my head, that it only belonged if it was something relevant to their story.
>
> (Pitzer, 2010)

Thus, in a desire to be transparent about her interventions in the Lacks's lives, Skloot includes herself in the role of investigating journalist, but strictly delimits the portrayal of this character in a way that she does not the characterisations of the Lackses.

Narrative priorities and ethical elisions

Despite holding back on details about her own life where it does not serve the aim of her book, Skloot offers considerable detail about the Lackses' lives in the service of narrative immersion. This has raised concerns about the ways in which the book configures issues around race. Among the charges are that the book reinforces harmful stereotypes. For instance, the physician and chair of the Tuskegee Syphilis Study Legacy Committee, Vanessa Northington Gamble, argues that the book: "at times depicts a stereotypically dysfunctional black family. Skloot spotlights promiscuity, mental illness and incarceration without any explanation for why these are vital to her larger narrative" (2014). Gamble does not enter into specifics in her brief review, but much of the book's discussion of Deborah's brother Zakariyya – who spent time in jail and who struggled with emotional hardship – might serve as an example, as would the portrait of Deborah as, at times, emotionally unstable. In one scene, Skloot depicts Deborah as suffering from a breakout of hives due to the mounting stress she has experienced over the years. Skloot's description of the appearance of the hives and accompanying welts is vivid and unflattering (p. 286). In the scene that immediately precedes the appearance of the hives, Skloot describes Deborah – driven by the fear that Skloot is conspiring with Johns Hopkins University – attacking her and pinning her to the wall of her hotel room (p. 283). Neither scene contributes to the stated mission of the book. As Karla Holloway observes, characterisations such as these are part of a pattern of appropriation of women's bodies – and in particular, the bodies of Black women – in the service of "public spectacle". Private moments in the lives of Henrietta and Deborah become, says Holloway, "an open narrative for public consumption" (2012: 16:23).

The scene in which Deborah attacks Skloot may be traced to Skloot's desire, as a journalist, to be transparent about her own involvement in the Lacks family's story. Yet Lindsay Morton asks of this scene: "Is this disclosure transparency? Or is it sculpting a narrative? Can these two be separated? This example suggests not" (2014: p. 255). While Skloot offers no clear indication of how the scene serves the book's aims, she did, in an interview, explain her decision to include her own reactions in it. Her editor, she explained, felt that early drafts where she had endeavoured to remove every trace of herself were uncompelling and alienating (Pitzer, 2010). By this account, her final authorial judgement was based on narratival, rather than ethical, considerations.

The scenes with Deborah further the development of the figure, who becomes the most prominent character in Skloot's narrative. This focus on Deborah – and her relationship with Skloot – was retained in the 2017 TV film adaptation, directed by George C. Wolfe and starring Oprah Winfrey as Deborah. The highs and lows of the interactions between the two women undoubtedly make for a more compelling narrative, revealing the ways in which Skloot's journey was marked by difficult-to-navigate obstacles. Reflecting on the book, however, Skloot sees the obstacles that Deborah navigated, rather than her own journey, as the focus of the narrative: "for me, it was really the story of Deborah: her struggle to learn who her mother was, to come to terms with the cells. [...] to get to the place where she does in the end, where she's pretty happy about the cells" (Pitzer, 2010). While the book also foregrounds Skloot's own presence as a character, Skloot seeks to downplay the implications of this: "I'm only a vehicle to get to Deborah, to show where she is today and the impact that all of this has had on her" (Pitzer, 2010).

Because Deborah died near the end of Skloot's work on the book, Skloot tells Deborah's story posthumously, as she does Henrietta's. Deborah's prominence in the book gives her a voice after death – a page of verbatim transcript after the prologue is explicitly entitled "Deborah's Voice" – and Skloot presents her as an active participant in the book project. For instance, at one point she records Deborah's instructions for how the book should unfold: "There's two sides to the story, and that's what we want to bring out. Nothing about my mother is truth if it's about wantin to fry the researchers. It's not about punish the doctors or slander the hospital. I don't want that" (p. 250). Deborah appears committed – at times, at least – to sharing her family's experiences with the public at large.

Thus, Skloot defends her project as an act of ventriloquism. Some critics, however, have identified a more exploitative stance. bell hooks, for instance, characterises Skloot's treatment of Henrietta Lacks as a "colorful backdrop, subordinated to the story of HeLa" (2013: p. 91). hooks ascribes Skloot's motivation to that of a "science journalist who wants to reveal a story to the world that will bring her fame and glory" (hooks, 2013: p. 90). Skloot, hooks contends, does not recognise that a white author cannot avoid bias nor fully comprehend how "systems of imperialist white supremacist capitalist patriarchy have worked together – historically in Africa and in the United States from slavery to the present day – to deny life to the black female body" (p. 90).

The question of access to healthcare shows how Skloot's approach overlooks systemic problems. Skloot pays due attention to the disparity between the renown for Henrietta's cells and the lack of healthcare that many of Henrietta's relatives and descendants experience. Yet, as Frances Gabbay puts it, the fundamental moral challenge here is that: "an American family – any American family – cannot afford health care" (2012: p. 118). Priscilla Wald speaks to the same problem, but expands the scale of the issue when she points out that: "Henrietta Lacks should not have been on the ward because such a segregated space should never have existed" (2012: pp. 187-188). Skloot's focus on the extraordinary qualities of the cells, the degree to which that distinguishes Henrietta's legacy and the plight of the Lacks family, thus results in a missed opportunity to engage with broader issues of inequality when they do not immediately serve her goal of narrating her journey with Deborah in uncovering the history of Henrietta and the HeLa cell line.

Skloot's discussion of the 'contamination' controversy that surrounded the cell line illustrates how the combination of a narrative approach with science journalism can curtail critical insights into the issues being reported. Skloot devotes a significant amount of space to the scientific debate and the crisis that the incredible robustness of the HeLa cells initiates. That robustness led to the cells being found in other cell lines, raising concerns that researchers around the world were not necessarily working with the type of cells they had thought they were working with. Skloot uses this history as an occasion to convey both how widespread (and, therefore, important) the HeLa cell line is and how hardy the cells themselves are. However, she fails to engage with the degree to which connotations of race entered the conversation about the so-called 'contamination'. As Hannah Landecker has identified, discussions of the controversy have been structured by "metaphors of miscegenation" (2007: p. 169). Lisa Weasel (2004) similarly locates propositions that the cell line be considered a new species in the context of the troubled histories of equating race, species and intelligence. Such insights require a step back from the events themselves. To report them would be to break the narrative immersion that *Immortal Life* offers and which makes the book so compelling. This is the ethical dilemma that the narrative approach engenders and that is compounded in a narrative about science. Even in the Afterword, when Skloot does pull back from the story world to provide a more distant perspective, the preceding narrative determines that she focuses on the issues that the on-going scientific research raises: the legal and ethical issues around consent rather than the racial prejudice and systemic inequalities that exacerbate these issues.

Despite the book's failure to articulate structural inequalities for what they are, its exposure of the Lacks family's mistreatment at the hands of the medical research community ensured that the book was frequently cited in debates about medical ethics, especially around informed consent and the continued use of the HeLa cell line. Skloot reports that she and the Lackses spoke on these issues to audiences "by the thousands" in the three years after *Immortal Life* was first published (Skloot, 2013). Skloot's authorship had afforded her the authority to intervene and to do so with significant impacts. In 2013, for instance, she wrote in *The New York Times* of her concern that recent developments in research with HeLa cells had occurred without the notification of – let alone consent from – the Lacks family. Research teams in Europe had just completed a full sequencing of the HeLa genome, with one team's research published already and another's forthcoming at the time. Skloot argued strongly against the assumption that anonymising data erases privacy concerns. As evidence, she cited one scientist who "uploaded HeLa's genome to a public Web site called SNPedia, a Wikipedia-like site for translating genetic information. Minutes later, it produced a full report of personal information about Henrietta Lacks, and her family" (Skloot, 2013).

Skloot served as both an ethical commentator and gatekeeper during this episode, putting the researchers in contact with the family as well as airing her views in the media (Calloway, 2013a). The family met with the director of the US National Institutes of Health, "with help from Skloot", resulting in an agreement that the sequenced genome would be released under a controlled access arrangement (Hudson and Collins, 2013). Access to the sequence was to be decided by a board comprised of scientists, bioethics experts and two members of the Lacks family.

In another extra-textual intervention – one that Skloot promised to Deborah Lacks in an exchange documented in the book (p. 251) – Skloot set up a charity, the Henrietta Lacks Foundation, to provide support for education and healthcare to individuals (and their descendants) who had been subject to research without their knowledge or consent. Initially funded by proceeds from the book and the film adaptation, in recent years the Foundation has received donations from medical research organisations seeking to offer some financial compensation for their use of HeLa cells (*Nature*, 2020).

It is rare for narrative journalism to have real-life impacts beyond selling books or moving readers. As a consequence of prioritising narrative immersion through subjective narration, Skloot was unable to account for the structural inequalities around which her narrative revolved, the explication of which required a more objective and distant perspective than an immersive narrative can offer. Yet despite this, the commercial success that Skloot's narrative strategy secured, ensured that she had a platform from which to call for policy changes and to help the Lacks family be heard in debates about the uses of HeLa cells.

Conclusion

Tim Radford, former science editor at the *Guardian* newspaper, notes that while an article in a scholarly journal can make an impact even if no one reads it as long as the article's abstract makes the rounds in the right academic circles, if a newspaper article goes unread, then "it might as well have had no existence at all" (2008: p. 98). "Newspapers", Radford writes, "exist to be read." The same applies to book-length journalism. In a review of *Immortal Life*, Robert Gifford reports that, while working on the book, Skloot would often have to fend off the dismissive challenge: "Who wants to read a book about cells?" (2012: p. 103). The book's countless readers have since proved those early challengers wrong: they did very much want to read *this* book about cells.

The commercial success of the book confirms that Skloot was right to believe that her subject was compelling. In addition to sales, awards and a place on many curricula, the book has attracted considerable media attention and has been cited in calls for policy changes. Yet there is no consensus on whether *The Immortal Life of Henrietta Lacks* is ultimately in the service of Skloot's subject or of herself, or on where the book falls in its effort to perform advocacy and to do so ethically. Some public health researchers, such as Lydia Hunt and her co-authors (2020) or Lisa Sullivan and Sandra Galea (2019), see the book as a productive pedagogical resource for exposing healthcare disparities; others, such as Christine Vogt-William, argue that it is a work of "textual colonization" (2017: p. 142).

There is also something problematic about the book's centrality in the continued discussions about both HeLa and the Lacks family: as if, having reluctantly decided to write herself into the story, Skloot cannot be written out of it again. For instance, even a decade after *Immortal Life* was first published, the *New York Times* still included a brief reference to the book in a news report about a settlement between the Lacks family and a biotechnology company (Holpuch, 2023). For comparison, sociologist Hannah Landecker – who has written extensively on the cultural, political and bioethical significance of HeLa and Henrietta Lacks (and whose work Skloot cites) – very much contributes to the lasting conversation on the subject, but with far less of a personal imprint on it than Skloot. And yet, without Skloot, that conversation would be significantly lesser in scale.

Perhaps all writers who seek to expose injustice or report on societal ills, despite their best intentions, inevitably enlist themselves in a Faustian bargain, wherein the more tragic the issue on which they report, the more likely they may be to gain recognition for their work and the harder it then is to extricate themselves from the legacy of their subject. As Skloot's book makes clear, the choice to use narrative conventions propels the journalist into an ethical minefield. Yet the advocacy in which Skloot's book engages and the extra-textual work Skloot performs or facilitates in support of the Lacks family would not be possible for a reporter who had to maintain an objective distance.

Narrative journalism, by encouraging readers to relate to characters, can attract larger and more invested audiences than traditional journalism. This is a challenge for science journalism, where the subjects are inanimate objects, abstract concepts or, at best, characters with privileged access to knowledge. As other science journalists have done in long-form writing, Skloot achieved narrative immersion by prioritising character over explanation. Yet she also sought to draw attention to an ethical injustice by switching the focus of the story from scientists to those affected by the science. This, in turn, demanded a further step. To be truthful about her own role in the story, Skloot had to insert herself into the narrative. This authorial transparency, adopted for the sake of ethical journalism, has the simultaneously ethically damaging countereffect of limiting the narrational perspective. As a narrative strategy, rather than an ethical strategy, it was impactful, serving to secure the connection between reader and author and thus ensuring that the story was widely accessible and well-positioned to effect change. It was also a commercial success. As Skloot became a character in her subjects' stories, their stories became hers – as it turned out, a highly profitable appropriation.

The science journalist needs to work harder than most to attract readers. Narrative journalism provides an effective means for accomplishing this, yielding a level of immersion that is difficult to achieve with expository writing. Yet an immersive approach renders ethics subservient to narrative. This is the Faustian bargain on which narrative journalism is based.

References

Booth, W.C., 1988. *The Company We Keep: An Ethics of Fiction.* Berkeley: University of California Press.

Calloway, E., 2013a. HeLa publication brews bioethical storm. *Nature*, 27 Mar., https://doi.org/10.1038/nature.2013.12689.

Craig, D., 2006. *The Ethics of the Story: Using Narrative Techniques Responsibly in Journalism.* Lanham: Rowman & Littlefield.

Eason, D., 1984. The new journalism and the image-world: Two modes of organizing experience. *Critical Studies in Mass Communication*, 1(1), pp. 51–65.

Gabbay, F.H., 2012. An American woman and the right to health. *Psychiatry*, 75(2), pp. 113–119.

Gamble, V.N., 2014. *The Immortal Life of Henrietta Lacks* reconsidered. *The Hastings Center Report*, 44(1), https://doi.org/10.1002/hast.239.

Gifford, R.K., 2012. A book about cells and much more. *Psychiatry*, 75(2), pp. 103–106.

Giorgis, H., 2017. Who owes the Henrietta Lacks family? *The Ringer*, 24 Apr., https://theringer.com/immortal-life-henrietta-lacks-hbo-film-oprah-winfrey-rebecca-skloot-82a1e94009f3.

Holloway, K., 2012. *Henrietta Lacks and the Ethics of Privacy*. Georgia Tech Library, http://hdl.handle.net/1853/45282.

Holpuch, A., 2023. Family of Henrietta Lacks settles with biotech company that used her cells. *The New York Times*, 1 Aug., https://www.nytimes.com/2023/08/01/science/henrietta-lacks-cells-lawsuit-settlement.html.

hooks, b, 2013. *Writing beyond Race: Living Theory and Practice*. New York: Routledge.

Hudson, K. and Collins, F., 2013. Family matters. *Nature*, 500, pp. 141–142, 7 Aug., https://doi.org/10.1038/500141a.

Hunt, L., Tkach, N. and Scott, L.B., 2020. Analysis of an interprofessional experiential learning program utilizing the case of Henrietta Lacks. *Pedagogy in Health Promotion*, 6(3), pp. 203–211.

Landecker, H., 2007. *Culturing Life*. Cambridge: Harvard University Press.

Lasso, R.A., 2011. The Immortal Life of Henrietta Lacks. *JAMA*, 305(11), pp. 1143–1144.

Malcolm, J., 1990. The morality of journalism. *New York Review of Books*, 1 Mar., pp. 19–23.

Morton, L., 2014. Evaluating the effects of epistemic location in advocatory literary journalism. *Journalism*, 17(2), pp. 244–259.

Nature, 2020. Editorial: Henrietta lacks: Science must right a historical wrong. *Nature*, 585(7823), p. 7. https://doi.org/10.1038/d41586-020-02494-z.

Nisbet, M.C. and Fahy, D., 2013. Bioethics in popular science: Evaluating the media impact of *The Immortal Life of Henrietta Lacks* on the biobank debate. *BMC Medical Ethics*, 14(1), pp. 1–9.

Phillips, L.A., 2019. From major to minor: Literary journalism and the first person. In: Dow, W.E. and Maguire, R.S. (eds.), *Routledge Companion to American Literary Journalism*. New York: Routledge, pp. 385–395.

Pitzer, A., 2010. Rebecca Skloot on narrating history: "looking for that one family, that one person, that one moment that will help hold everything together". *Nieman Storyboard*, 16 Jul., https://niemanstoryboard.org/stories/rebecca-skloot-immortal-life-of-henrietta-lacks-interview-narrative/.

Radford, T., 2008. Scheherazade: Telling stories, not educating people. In: Bauer, M.W. and Bucchi, M. (eds.), *Journalism, Science and Society*. New York: Routledge, pp. 95–100.

Ricketson, M., 2021. Literary journalism and global media ethics. In: Ward, S.J.A. (ed.), *Handbook of Global Media Ethics*. Cham: Springer, pp. 743–761.

Roston, E., 2010. Book review: 'The Immortal Life of Henrietta Lacks' by Rebecca Skloot, 31 Jan., http://www.washingtonpost.com/wp-dyn/content/article/2010/01/29/AR2010012902147.html.

Seaman, D., 2009. The Immortal Life of Henrietta Lacks. *Booklist*, 106(7), p. 18.

Skloot, R., n.d. FAQ: On writing *The Immortal Life of Henrietta Lacks. Rebecca Skloot*, https://rebeccaskloot.com/faq/#questions-writingimmortal.

Skloot, R., 2010. *The Immortal Life of Henrietta Lacks*. New York: Crown Publishers.

Skloot, R., 2013. The Immortal Life of Henrietta Lacks, the Sequel. *The New York Times*, 23 Mar., http://www.nytimes.com/2013/03/24/opinion/sunday/the-immortal-life-of-henrietta-lacks-the-sequel.html.

Sullivan, L.M. and Galea, S., 2019. *Teaching Public Health*. Baltimore: Johns Hopkins University Press.

van Krieken, K. and Sanders, J., 2021. What is narrative journalism? A systematic review and an empirical agenda. *Journalism*, 22(6), pp. 1393–1412.

Vogt-William, C., 2017. HeLa and the help: Justice and African-American women in white women's narratives. In: Bartels, A. et al. (eds.), *Postcolonial Justice*. Leiden: Brill Academic Publishers, pp. 141–176.

Wald, P., 2012. American Studies and the politics of life. *American Quarterly*, 64(2), pp. 185–204.

Weasel, L.H., 2004. Feminist intersections in science: Race, gender and sexuality through the microscope. *Hypatia*, 19(1), pp. 183–193.

Wolfe, T. and Johnson, E.W. (eds.), 1973. *The New Journalism* (1st ed.). New York: Harper & Row.

CONCLUSION

A conversation with the future

Felicity Mellor

In one of the first studies of science journalism as newsroom practice, Anders Hansen suggested that the key to understanding science journalists is "to recognize that they are, in their practices and professional beliefs, journalists first and specialists second" (1994: p. 111). There may be few other beats where this would need saying. As the contributions in this book have shown, science journalism has repeatedly had to assert its independence from science, even as it uses standard journalistic techniques to reach public audiences. Science journalists' practices align with those of other journalists in their commitment to journalistic norms, news values and storytelling (Chapters 8 to 10); yet in their dependence on the community on whom they report, science journalists are journalistic outliers (Chapters 4 to 7). This tension was present at the origins of the profession when the first specialist science journalists, backed by the institutions of science and finding a collective voice through professional organisation, made a case for their presence in the newsroom on the basis of the social value of science (Chapters 1 to 3).

As a journalistic field, the future of science journalism is subject to the same pressures as other branches of journalism: disruptive media technologies, a collapsing business model, job cuts, information saturation, audience fragmentation and competition from non-journalistic media producers. As a specialism, science journalism faces additional pressures stemming from the increasing capacity and proficiency of science PR, but may also be afforded some protection through its alliances with science.

While this book was in preparation, generative artificial intelligence (AI) broke out of the confines of science and technology reporting to become a major news story. The public release of the ChatGPT programme showcased the technology's ability to write coherent natural language responses to any

DOI: 10.4324/9781003326724-15

prompt. By this time, it was already clear that such tools will not only provide material for journalists to report on, but will have significant impacts on how journalists work. In a 2023 survey of newsroom executives, half of the respondents reported that their newsrooms were already using tools such as ChatGPT, albeit on a limited basis, and over 80% believed that newsroom roles and responsibilities would change as a result of the increasing use of such software (WAN-IFRA, 2023). The most common use was to generate text summaries or lists (AI has been used to distil sports results and financial information for some years), but the survey found that the technology was also being used for research, to correct text and, in a third of newsrooms, to produce articles. For some newsrooms, this last is a step too far; for instance, the UK's *Guardian* newspaper has a policy forbidding the inclusion of AI-generated text in its published content, citing concerns that "it is not reliable or consistent, and tends to introduce errors and inaccuracies in unpredictable ways" (GNM, 2023: p. 12). Other outlets, however, have based their business models on automated content creation; for example, a US company specialising in well-being and astrology websites claims to have used the technology to create thousands of articles in the course of a year (Guaglione, 2023).

Of course, science journalism demands a level of veracity that most science journalists would not associate with astrology; for science journalism, therefore, the stakes are considerably higher. In one of the first experiments with AI-generated content, the magazine *Men's Journal* published an article about low testosterone. The piece included medical claims and lifestyle advice that, according to one expert in the field, sounded true but were in fact full of inaccuracies. In defending itself, the magazine noted that the AI had drawn on content originally written by journalists and that this was where the errors had originated (Christian, 2023). AI's mistakes, the magazine implied, are ultimately human mistakes.

To explore how the AI-driven newsroom will impact on science journalism, I asked the opinion of a key player helping drive through these changes: ChatGPT. The following is a transcript of our conversation, lightly edited to avoid repetition.

FM: What is the future of science journalism?

ChatGPT: The future of science journalism will likely involve more multimedia content, data-driven storytelling, and increased collaboration between scientists and journalists. It'll also adapt to emerging technologies and prioritize accurate, accessible, and engaging communication of complex scientific topics to a broader audience.

FM: So more of the same then? Will nothing change?

ChatGPT: While some elements of science journalism will remain familiar, significant changes are expected. Advancements in technology, like virtual reality and AI, may revolutionize how stories are told. Additionally, a focus on transparency, ethics, and addressing misinformation will likely be prominent in the evolving landscape.

FM: The use of AI seems particularly significant. Tell me how generative AI is already being used by science journalists.

ChatGPT: It's used to generate automated news reports from data, create summaries of scientific papers, and even aid in fact-checking. Natural language generation models like GPT-3 can help streamline content creation and make complex research more accessible to a broader audience.

FM: What additional uses are we likely to see in the near future?

ChatGPT: We can expect generative AI to play a more significant role in science journalism. This may include personalized content recommendations for readers, real-time language translation for global coverage, and AI-generated interactive simulations or visualizations to explain complex scientific concepts. It can also help identify emerging trends and research topics, making journalism more proactive.

FM: This is all very positive. Don't you see any downsides?

ChatGPT: Indeed, there are potential downsides. One concern is the risk of AI-generated content being manipulated or used to spread misinformation. Additionally, heavy reliance on automation may reduce the need for human journalists, impacting employment in the field. Ethical issues related to AI's decision-making processes and biases also need to be carefully addressed. Striking a balance between AI assistance and responsible journalism will be crucial.

FM: We may soon face a situation where a scientific paper written by AI will be promoted in a press release written by AI, which will then be selected for coverage at a news outlet by AI, where it will be prepared for publication by AI and – perhaps after translation by AI – will be published by AI; whereupon AI in the form of social media algorithms will draw it to readers' attention, readers will turn to AI to help them judge the credibility of what they are reading and online bots will comment on it. Ever more energy will be burnt up enabling AI to talk to AI. What is the point?

ChatGPT: The scenario you describe highlights the increasing role of AI in the entire information ecosystem. While AI can enhance efficiency and accessibility, it also raises concerns about the loss of human touch, editorial oversight, and potential bias. The point should be to ensure responsible AI use, maintain human editorial control, and uphold journalistic values, ethics, and credibility amid this evolving landscape. Striking a balance between AI's benefits and human stewardship is key.

FM: And if we don't manage that? What is the worst case scenario you can imagine?

ChatGPT: The worst-case scenario in an AI-dominated information ecosystem could involve unchecked misinformation, algorithmic bias amplifying divisive content, a decline in journalistic quality, loss of jobs in the industry, and a diminished public trust in media. This could lead to a fractured, polarized society with diminished access to reliable information, potentially undermining democracy and societal cohesion. Responsible AI governance and ethical journalism practices are vital to avoid such a future.

FM: Yet you didn't mention any of that in your first answer. I wonder why not?

ChatGPT: I apologize for any oversight in my initial response. In providing concise answers, I may not have covered all aspects of the question. It's important to address both the positive and negative potential outcomes of AI in science journalism to provide a comprehensive view of the topic. Thank you for pointing out the omission.

In a preliminary conversation, I had made a similar point. ChatGPT had thanked me that time too, promising to learn from its omission. That it did not reminds us that a narrow techno-optimism is engineered into the technologies that are about to transform global society: the need for critical science journalism, produced by thinking humans, will be greater than ever.

References

Christian, J., 2023. Magazine publishes serious errors in first AI-generated health article. *Neoscope*, 18 Feb., https://futurism.com/neoscope/magazine-mens-journal-errors-ai-health-article.

GNM, 2023. *Editorial Code of Practice and Guidance*. Guardian News & Media, https://www.theguardian.com/about/2023/jul/07/guardian-editorial-code-of-practice-and-guidance-2023.

Guaglione, S., 2023. How one publisher is using generative AI to publish thousands of evergreen posts, create a chatbot. *Digiday*, 29 Mar., https://digiday.com/media/how-one-publisher-is-using-generative-ai-to-publish-thousands-of-evergreen-posts-create-a-chatbot/.

Hansen, A., 1994. Journalistic practices and science reporting in the British press. *Public Understanding of Science*, 3(2), pp. 111–134.

WAN-IFRA, 2023. *Gauging Generative AI's Impact on Newsrooms*. World Association of News Publishers, https://wan-ifra.org/2023/05/new-genai-survey/.

INDEX

Note: Locators in *italics* represent figures; **bold** indicate tables in the text and page numbers followed by 'n' refer to notes.